C000127780

PEN...
HINDUIS...

A historian, environmentalist and writer based in Chennai, Nanditha Krishna has a PhD in ancient Indian culture from the University of Bombay. She has been a professor and research guide for the PhD programme of C.P.R. Institute of Indological Research, affiliated to the University of Madras. She was the honorary director of the C.P. Ramaswami Aiyar Foundation from 1981 and was elected president in 2013. She is the founder-director of its constituents, including C.P.R. Institute of Indological Research, C.P.R. Environmental Education Centre, C.P. Art Centre and Kanchi Museum of Folk Art. She is the author of several books, including *Sacred Plants of India*, *Sacred Animals of India*, *Book of Demons* and *Book of Vishnu* (Penguin India); *Madras Then*, *Chennai Now*, *Balaji Venkateshwara*, *Ganesha*, *Painted Manuscripts of the Sarasvati Mahal Library*; and *The Arts and Crafts of Tamilnadu* and *The Art and Iconography of Vishnu-Narayana*, among many others, besides numerous research papers and newspaper articles.

PRAISE FOR *SACRED PLANTS OF INDIA*

'This is a fascinating account of tree and plant worship in India from time immemorial. Worship of trees, some of which were believed to be home to spirits good and bad, was probably the oldest form of worship in India'—*Business Standard*

'*Sacred Plants of India* connects the dots between flora, mythology and science beautifully, tracing the socio-cultural-scientific roots of plant worship. It is not a coffee table book brightened up with colour palettes and is not for light reading. But it makes for an interesting, rewarding and inspiring read and will be a valuable intellectual resource for the future'—*Week*

'Will be of interest not only to environmentalists and conservationists but also nature lovers and those interested in having a small house garden. It is also meant for policy-makers and the common people to encourage greater participation in forest conservation. The detailed and inclusive research done for the book is indeed praiseworthy'
—*TerraGreen*, TERI

PRAISE FOR *SACRED ANIMALS OF INDIA*

'A good read at a time when the world as we know it is in the throes of an increasingly contentious debate on the future of our environment'
—*Mail Today*

'The book, written in simple straightforward language, treats the complex subject with the confidence that is born out of meticulous and thorough research and strong convictions. Tribal lore, folklore, ancient scriptures, traditional tales, history, scriptural texts, cave paintings—no source seems to have been left uncombed in the effort to chronicle the history and progress of the Indians' attitude towards animals'—*Book Review Literary Trust*

'Each and every animal, however insignificant in other eyes, attracts the attention of a devout Hindu. He in his own way discovers some mystique in it, which compels him to bow his head with a sense of worship'—*Dawn*

nanditha krishna

HINDUISM
and
NATURE

PENGUIN BOOKS
An imprint of Penguin Random House

PENGUIN BOOKS

USA | Canada | UK | Ireland | Australia
New Zealand | India | South Africa | China

Penguin Books is part of the Penguin Random House group of companies
whose addresses can be found at global.penguinrandomhouse.com

Published by Penguin Random House India Pvt. Ltd
7th Floor, Infinity Tower C, DLF Cyber City,
Gurgaon 122 002, Haryana, India

First published in Penguin Books by Penguin Random House India 2017

The views and opinions expressed in this book are the author's own and the
facts are as reported by her which have been verified to the extent possible,
and the publishers are not in any way liable for the same.

ISBN 9780143427834

Typeset in Bembo Std by Manipal Digital Systems, Manipal
Printed at Replika Press Pvt. Ltd, India

Aum dyauh shantir antariksham shantih
Prithivi shantir aapah shantir aushadhayah shantih
Vanaspatayah shantir vishvedevah shantir brahma shantih
Sarvam shantih shantireva shantih
Sa ma shantiredhi
Aum shanti shanti shantih

Yajur Veda Samhita (36:17)

May peace radiate in the whole sky and in
the vast ethereal space,
May peace reign all over this earth, in water,
in all herbs and the forests,
May peace flow over the whole universe,
May peace be in the Supreme Being,
May peace exist in all creation, and peace alone,
May peace flow into us.
Aum—peace, peace and peace!

Contents

1

Introduction

So long as the earth is able to maintain mountains, forests
and trees
Until then the human race and its progeny will be able to
survive.

—*Durga Saptashati*, 'Devi Kavacham', 54

The Indian housewife starts her day by cleaning the space outside the front door and decorating it with beautiful designs made of rice flour. Apart from beautifying her home, she is also feeding the ants and does not need to spray insecticide to keep them out.

When she bathes, she prays that the water may be as sacred as the River Ganga, which has proven antimicrobial qualities.

She encircles the pipal tree seven times in a ritual binding the Indus, Vedic, Hindu, Buddhist, Jain and tribal traditions. It is a unique tree which filters impurities in the air and releases oxygen day and night.

She pours water over the sacred basil plant—tulsi—in the centre of her house, for it prevents coughs, colds and fevers.

She places a small bowl of cooked rice on the roof for her departed ancestors, which is eaten by crows who keep the outer environment clean.

She will sweep her house only under bright daylight, for she fears that she may harm or dislodge small insects from their homes in nooks and crannies if she sweeps in the twilight or darkness.

Every aspect of her life is intimately connected with nature and the environment, and scientific environmental management. Unfortunately, all that was good and preserved in the name of culture and tradition has been discarded in the name of modernization and development.

The basis of Hindu, Buddhist and Jain culture is dharma or righteousness, incorporating duty, cosmic law and justice. It is *sanatana*, or eternal, for it is without beginning or end, and it supports the whole universe. Every person must act for the general welfare of the earth, humanity, all creation and all aspects of life: 'Dharma is meant for the well being of all living creatures. Hence that by which the welfare of all living creatures is sustained, that for sure is dharma' (Mahabharata, XII.109.10). Dharma means many things: righteousness, duty, justice and law. Every divine incarnation is born to restore it.

The verses of the Vedas express a deep sense of communion of man with god. Nature is a friend, revered as a mother, obeyed as a father and nurtured as a beloved child. It is sacred because man depends on it and because of this everything is sanctified, including man and the terrifying aspects of nature, such as landslides, earthquakes and storms. Natural phenomena are the manifestations or expressions of the gods and not the gods themselves. They express the principles that govern the world and the cosmic order, *rita*.[1]

In the *Rig Veda*, Vritra is a serpent or dragon called Ahi, the personification of drought and an adversary of Indra, the god of rain and thunder. Vritra keeps the waters captive until he is killed by Indra, who destroys all his ninety-nine fortresses and liberates the imprisoned rivers. At Indra's request, Vishnu, the god of sun, strides across the firmament, the sun and the rain combining to destroy Vritra (I.6.1; II.22.1; VI.20.2). While Indra does the actual destroying, Vishnu is his friend and helper.

The word Vritra comes from the root '*vri*', meaning to enclose, and is used for a harasser or foe. Vritra is the personification of the evil power of drought. Vishnu's assistance to Indra may thus be interpreted as the help given by the god of sun to the god of rain when the latter is about to destroy the demon who prevents the waters from fertilizing the soil. The Vedic religion was, essentially, nature worship. Indra calls for the sun's aid to fight the demon of drought and to convert the clouds into rain. The war against Vritra was apparently an anthropomorphic representation of the diverse forces of nature which these gods represented.[2] It is not possible to agree with D.D. Kosambi's theory that Indra's war against Vritra is symbolic of the destruction of the dams and barrages built on the Indus by the Harappans.[3] We now know that the Vedic and Harappan cultures ran simultaneously,[4] therefore the theory of the destruction of one by the other is not tenable. Vedic religion was pantheistic, celebrating nature as divinity.

In Vedic literature, all of nature was, in some way, divine, part of an indivisible life force uniting the world of humans, animals and plants. The Vedas are dedicated to a variety of pantheistic deities called devas or the Shining Ones, representing the stars in the firmament and forces beyond human knowledge or control. Chief among them was Indra. Soma was a sacred plant and Agni, the divine fire. There are several solar deities: Surya, Savitr and Aditya, while Ushas was the dawn, a female deity. Vishnu was also a solar deity, symbolized by his three

steps across the firmament: the morning and evening sun and
the midday orb. Pushan represented agriculture. Dyauspitr
was the divine father (the sky, father of the heavens), Prithvi
was Mother Earth and Vayu, the wind. The Rivers Sarasvati,
Sindhu (Indus) and the latter's tributaries—Shutudri, Parushni,
Ashkini, Vitasta and Vipasa or the Sapta Sindhava—were all
regarded as sacred. Thus the concept of the sacred environment
was established in the Vedic period itself.

Vedic people were one with nature. 'One is that which
manifests in all' (Rig Veda, I.164.46) meant that everything is
related to everything else. Man had to recognize what powers
of nature he could not control and was thus compelled to
resort to prayer to win the cooperation of the winds and rains
to ensure the regularity of the monsoon; for the control of
earthquakes, forest fires and all major elements of nature. He
who sought nature's laws was rewarded: 'Those ancient sages,
our ancestors, observant of truth, rejoicing together with the
gods, discovered the hidden light, and, reciters of sincere
prayers, they generated the dawn' (Rig Veda, VII.76.4).[5]

> Ancient Indian theistic systems share a substantial
> commonality of conceptions concerning nature, seen as
> a phenomenal reflection of God's essence. The forms of
> nature are regarded as rooted in God, the transcendental
> Creator, and the various types of existence, phenomenally
> produced by nature, are seen as multifarious reflections of
> divine qualities. The beings of the world are considered
> as emanations from the transcendental unity of God, from
> where they come, and whence they are due to return.[6]

Nature or prakriti means 'making or placing before or at first,
the original or natural form or condition of anything, original
or primary substance'.[7] Formulated by the Samkhya school of

philosophy, it refers to primal matter with three different innate qualities (guna)—rajas (creative activity), sattva (calmness of preservation) and tamas (destruction)—whose equilibrium is the basis of all observed reality. While purusha is the masculine aspect of creation, prakriti is the feminine aspect of existence. Prakriti is dynamic, causing change, the 'primal motive force', an essential part of the universe and the basis of all creation.

Who created the world? What caused it to be? The Hindu concept of creation is presented differently in literature from about 2500 to 500 BCE. The Vedic, Upanishadic and Puranic literature and the epics describe the creator and the order of the trinity differently, but a single thought flows through them— that god and creation are the same. The Hymn of Creation (*Rig Veda*, 'Nasadiya Sukta', X.129) ('*naasat*' meaning not the non-existent) on cosmology and the origin of the universe, given below, has been described by the astronomer Carl Sagan as India's 'tradition of skeptical questioning and unselfconscious humility before the great cosmic mysteries'.

Then there was neither non-existence nor existence,
Then there was neither space, nor the sky beyond.
What covered it? Where was it? What sheltered it?
Was there water, in depths unfathomed?

Then there was neither death nor immortality
Nor was there then the division between night and day.
That One breathed, breathlessly and self-sustaining.
There was that One then, and there was no other.

In the beginning there was only darkness, veiled in darkness,
In profound darkness, a water without light.
All that existed then was void and formless,
That One arose at last, born of the power of heat.

In the beginning arose desire,
That primal seed, born of the mind.
The sages who searched their hearts with wisdom,
Discovered the link of the existent to the non-existent.

And they stretched their cord of vision across the void,
What was above? What was below?
Then seeds were sown and mighty power arose,
Below was strength, Above was impulse.

Who really knows? And who can say?
Whence did it all come? And how did creation happen?
The gods themselves are later than creation,
So who knows truly whence this great creation sprang?

Who knows whence this creation had its origin?
He, whether He fashioned it or whether He did not,
He, who surveys it all from highest heaven,
He knows—or maybe even He knows not.

Five thousand years ago, the sages of the *Rig Veda* showed a
clear appreciation of the natural world and its ecology, the
importance of the environment and the management of natural
resources (I.115, VII.99 and X.125). The *Rig Veda* dedicates a
whole hymn to the rivers ('Nadistuti Sukta'), while the hymn
to the earth ('Prithvi Sukta', Book 12) of the *Atharva Veda*
consists of sixty-three stanzas in praise of Mother Earth and
nature, and human dependence on the earth.

The *Rig Veda* goes on to say that the Vedas and the
universal laws of nature which control the universe and govern
the cycles of creation and dissolution were made manifest by
the All-knowing One. His great power produced the clouds
and the vapours, followed by a period of darkness, after which

the Great Lord and Controller of the Universe arranged the motions to produce days, nights and other durations of time. The Great One then produced the sun, the moon, the earth and all other regions as He did in the previous cycles of creation (X.190.1–3).

'In the beginning there was the Self alone. He transformed himself into man and woman. Later, He transformed Himself into other creatures: bipeds and quadrupeds. In this way He created everything that exists on earth, in water, and sky. He realized: "I indeed am creation, for I produced all this." Thence arose creation' (*Brihadaranyaka Upanishad*, 1–5).

And, in the *Bhagavad Gita* (7.6), Lord Krishna says to Arjuna:

I am the source of all spiritual and material worlds. Everything emanates from Me . . .

I am inexhaustible time, and of creators I am the Supreme Creator . . .

I am all-devouring death, and I am the generating principle of all that is yet to be . . .

The earliest Sanskrit texts, Vedas and Upanishads, have preached about the non-dualism of the Supreme Power that existed before Creation. God as the efficient cause and nature (prakriti) as the material cause of the universe is unconditionally accepted, as is their harmonious relationship. However, they differ in their theories on the creation of the universe, which was answered in the *Rig Veda*.

God is called Brahma, the creator of the universe, Vishnu, the all-pervading preserver and Rudra, the punisher of the wicked. The idea is that 'God is one; Gods are differently named concepts of the One Being' (*Atharva Veda*, II.1.3). This primordial or cosmic matter of nature is made up of five elements—*prithvi* (earth), *vayu* (air), *agni* (fire or energy), *aapa* (water) and *akasha* (space)—better known as the *pancha-maha-bhuta*. Their proper balance and harmony are essential for the well-being of humankind, and maintenance of this harmony is a dharma, or righteous duty. The *Maitrayani Upanishad* has a beautiful analogy of Brahman as a tree with its roots above and its branches below, the branches being earth, water, air, fire and space. In terms of the human body, this is likened to the five senses: space is sound, air is touch, fire is colour, water is taste and earth is smell. Nature is thus an indivisible part of the existence of all beings. The earth and its inhabitants are part of a highly organized cosmic order called rita and any disruption results in a breakdown of peace and the natural balance.

The Tamil *Purananuru* (2.1–10) compares the Chera king to the pancha-maha-bhuta, as the ideal of perfection:

> Oh Lord of Glory! You possess
> Forbearance with enemies, extensive planning skill
> And valiant glories, resemble the patience of the Earth (*prithvi*),
> Dense with its packed dust-particles and
> Of the expanse of the Sky (*akasha*), high over the earth
> And of the might of the Wind (*vayu*) embracing the sky
> And of the destruction of Fire (*agni*) joined with wind
> And of the beneficence of Water (*aapa*) opposed to fire
> Oh Lord of the good land of ever fresh resources,
> Where the Sun, born on your sea, again
> Bathes by setting in your western sea of white-crested waves!

God and prakriti are one and the same, forming a single nucleus. Different elements are personified as parts of his body. He gave birth to purusha from a thousandth part of his body, a birth which was acknowledged by the rishis who called the newborn *manasa purusha*, who is eternal, infinite, ageless and immortal:

Father of all creatures, God, made the sky (*akasha*) first.
From the sky he made water (*aapa*) and from water he made
fire (*agni*) and air (*vayu*). From fire and air, earth (*prithvi*)
came into existence. Mountains are the bones, earth is the
flesh, sea is the blood, sky is the abdomen, rivers are nerves;
Air is His breath, Fire is His brilliance. The sun and moon
are the eyes of the Creator (Brahman). The sky is His head,
earth is His feet and the directions *(disha)* are his hands
(Mahabharata, *Mokshaparva*, 182.14–19).

The fusion of air, water and sky produced fire. These four
elements originate from the same source. When these four
elements moved downward, earth was produced. Later, these
five elements caused the birth of *srishti* (creation) and prakriti
(nature). The harmonious coexistence of these five elements is
essential for the well-being of life on earth.

'May this oblation, *jatavedas* (Agni), this reverence,
this praise, ever magnify you. Protect, Agni, our sons, our
grandsons and diligently defend our persons (bodies)' (*Rig
Veda*, X.4.7). Because body temperature is an indicator of
health or disease, Agni is eulogized as a physician and maker of
remedies (*Atharva Veda*, V.29.I). In his mysterious interactions
with life, Agni has also much to say about the functioning of
the mind: free the man possessed by the demon of disease (*Rig
Veda*, X.161, 162).

Agni and water are givers and sustainers of life, they are
'affectionate mothers . . . givers of all, givers of life' (*Rig Veda*,
IX.2). 'Waters, verily are medicinal; waters are the dissipaters
of disease; Waters are the medicines for everything; May they
act as medicine to you' (*Rig Veda*, X.137.6), 'Waters are the
most excellent . . . Agni is the most excellent . . . Earth declared
the third . . . praised is the lightning cloud' (*Rig Veda*, I.161.9).
'Leaders of the rains, you have caused the grass to grow upon

the high places; you have caused the waters to flow over the low places; for (the promotion of) good works. As you have reposed for a while in the dwelling of the inapprehensible (sun), so desist not today from (the discharge of) this' (*Rig Veda*, I.161.11). Soul and body are fire and water, the pair that enables life to go on (*Rig Veda*, X.11).

Throughout the Vedas, we sense a deep respect for life. Life is an important manifestation and expression of the gods, second perhaps only to Agni, the 'giver of life' (*Atharva Veda*, II.29). There would be no life without Agni. Thus the Vedic people recognized the presence of god in everything in the universe. The production of fire with ghee—by shedding water—represents the sacredness of the process and the need to follow rules. Trita, an ancient Vedic water deity living in remotest places, guides the priest Agni in the process (*Rig Veda*, X.115).[8]

The need to protect and conserve biological diversity is exemplified in the representation of the family and habitat of god Shiva, his consort Parvati and his two sons Karttikeya and Ganesha. His habitat is Mount Kailas, with snowy peaks representing the cosmic heavens. The crescent moon on his forehead denotes tranquillity; the constant stream of Ganga's water from the lock of hair on his head indicates the purity and importance of water; Nandi, his bull mount, represents the world of animals; serpents signify the presence of toxicity in nature; the lion used by his consort Parvati represents wildlife; the peacock, the mount of Karttikeya, represents the avian species; and the mouse, the mount of Ganesha, represents small underground animals. Different types of animals and birds inhabit the holy abode of Lord Shiva. Another significant aspect is the harmonious relationship between natural enemies. In Lord Shiva's household, various natural enemies live in harmony with each other. The carnivorous lion's food is the vegetarian bull, the peacock is the enemy of the serpent and the mouse is the serpent's food; nevertheless, all live together. Thus, when a devotee worships the family of Lord Shiva, he or she observes this coexistence and is influenced by what in contemporary times

Source: The C.P. Ramaswami Aiyar Foundation

Figure 1: The Shiva family by Raja Ravi Varma

might be seen as analogous to the concept of ecological harmony and respect for biological diversity[9] (Figure 1).

The Supreme Being or Brahman is the underlying power of unity, pervading all creation: forests and groves, trees and plants, animals, rivers, waterbodies, mountains, gardens, towns and precincts and seeds. Nature is venerated all over India. Every village has a sacred grove presided over by a local deity; every temple has a sacred garden and sacred tree; rivers and lakes are revered; and mountains are the dwelling place of the gods. Nature is a manifestation of the divine. In the *Bhagavad Gita* (4.7–8), Krishna says, 'I am the earth, I am the water, I am the air.'

A quotation from the *Atharva Veda* describes all forms of life. The hills and mountains covered with snow, the thick forests and the earth are sacred. It prays for the well-being and happiness of all living beings. It further craves for bountiful crops. May all tribes and nations prosper and may no one be subjected to your anger or suffer from natural calamities (XII.1.10). The world is one family—*vasudhaiva kutumbakam*—embracing all forms of life.

Another prayer addressed to Mother Earth asks for continuation of the blessings of the earth for all time to come. Let the people be blessed with a powerful intellect and the ability to converse with the gods. May the riches of the earth continue to be showered on the people of different speech, of diverse customs according to their homes, to ensure their material and spiritual wellness (XII.1.43–44). May Mother Earth, like a cosmic cow, give us thousandfold prosperity without any hesitation, without being outraged by our destructive actions (XII.1.43–45).

And finally, Mother Earth is acknowledged as the world itself: 'O Mother Earth! You are the universe and we are but your children. Grant us the ability to overcome our differences and live peacefully and in harmony, and let us be cordial and gracious in our relationship with other human beings' (*Atharva Veda*, XII.1.16). This *sukta* advises us to behave in a suitable manner towards nature and defines our duty towards the environment.

The *Atharva Veda* gives us a beautiful description of the relationship between human beings and nature: 'The earth, which possesses oceans, rivers and other sources of water and which gives us land to produce foodgrains and on which human beings depend on for their survival—may it grant us all our needs for eating and drinking: water, milk, grains and fruit' (XII.1.3).

Prajapati, the lord of creatures, is the creator and the protector of the sky, earth, oceans, people and animals. Humans have no authority over animals. Rather, they have duties and obligations towards all creation. The Hindu belief in the cycle of birth, death and rebirth requires Hindus to give all species equal respect and reverence, for they may be reborn as an animal, bird or insect in another life. The doctrine of ahimsa or non-violence is India's unique contribution to world philosophy. In Hinduism, Buddhism and Jainism, one's actions (karma) determine one's future life. Not only man, even God Himself has incarnated in several forms: Lord Vishnu's first four incarnations are as the fish (Matsya), tortoise (Kurma), boar (Varaha) and man-lion (Narasimha). He is also associated with several animals, such as Hanuman, the companion–devotee of Rama, and the cows, which are Krishna's companions, and thereby sacred. Every deity has a *vahana* (animal vehicle), who is his or her companion.

The most important aspect of Indian tradition is that of karma, associating all species with birth, death and rebirth. As the Supreme Being himself has been incarnated in several forms, people are advised to treat all species alike: 'One should look upon deer, camels, monkeys, donkeys, reptiles, birds and flies as though they were one's own children; what is that which distinguishes these from those?' (*Srimad Bhagavata Purana*, 7.14.9). One may be human in this life but may be reborn as an animal due to one's karma or actions (*Bhagavad Gita*, 14.15). Personal duty and actions, especially those that lead to goodness, are sattva. A life of action is rajas. And a wasteful or evil existence is tamas. Life is a journey through four ashramas or stages: *brahmacharya* (studenthood or celibacy), *grihasta* (householder), *vanaprastha* (preparing for renunciation) and, finally, *sanyasa* (renunciation) itself. Nature offers the right path leading to renunciation and moksha or liberation of the soul.

In Hinduism, there are no dos and don'ts, no god who sits in judgement. It is all cause and effect: one has to bear the consequences of one's behaviour, good or bad. God is kind and loving, not judgemental. People are responsible for their behaviour and one's karmas or actions lead to their own consequences in a future life.

There is a very strong and intimate relationship between the biophysical ecosystem and economic institutions. The two are inextricably held together by cultural relations. Hinduism has a definite code of environmental ethics. According to it, humans may not consider themselves above nature, nor can they claim to rule over other forms of life. Hence, traditionally, the Hindu attitude has been respectful towards nature.

> From Him too are the Gods produced manifold,
> The celestials, men, cattle, birds . . . (*Mundakopanishad*, 2.1.7)

> Those who are wise and humble treat equally the Brahmin,
>   cow, elephant, dog and dog-eater (*Bhagavad Gita*, 5.19)

Five thousand years ago, the sages of the *Atharva Veda* said, 'The earth's attributes are for everybody and no single group or nation has special authority over it' (XII.1.18). The hymn also describes the earth as the mother of all species living on it. 'Let the whole of humanity speak the language of peace and harmony and let all living beings live in accord with each other' (XII.1.16).

> *Karma* can . . . be considered the moral equivalent of the law of conservation of energy or the equivalence of action and reaction in the field of natural sciences. While it is true that what we are today is the result of our past deeds, it also follows that we are the makers of our future by the way we act at present. Thus, far from implying fatalism as is often wrongly believed, *karma* gives tremendous responsibility

to the individual and places in his own hands the key to his future destiny. Naturally, the unerring law of *karma* can work itself out only over a sufficiently long period of time; therefore the Hindu belief in reincarnation.[10]

Nature is the creation and manifestation of the Supreme Being. God is everywhere, says the *Bhagavad Gita* (13.13). Another reference in the *Srimad Bhagavata Mahapurana* (2.2.41) says that all the elements such as space, air, fire, water, earth, planets and even animals and plants, directions, trees, rivers and seas are but organs of god's body.

Pradushana (pollution) of any sort was abhorred: it was once a punishable offence. 'Punishment . . . should be awarded to those who throw dust and muddy water on the roads . . . A person who throws inside the city the carcass of animals . . . must be punished' (Kautilya, *Arthashastra*, 2.145). Environmental pollution (*vikriti*), a much-discussed problem of our times, was identified several millennia ago. 'From pollution two types of diseases occur in human beings. The first is related to the body and the other to the mind, and both are interrelated . . . coolness, warmth and air—these are three virtues of the body. When they are balanced in the body it is free from disease' (Mahabharata, XIII.16.811).

The great medical scientist Charaka was prescient when he predicted, 'Due to pollution of weather, several types of diseases will come up and they will ruin the country. Therefore, collect the medicinal plants before the beginning of terrible diseases and change in the nature of the earth' (*Charaka Samhita,* 'Vimanasthanam', 3.2). With the advent of modern medicine, Charaka has been forgotten.

India has a long tradition of conserving nature by giving it a spiritual dimension, but a fast-changing world, growing consumerism and population and the consequent pressure on land and natural resources has changed our value systems. The urgency of global warming and climate change calls for a greater response from the world's religions.

Hinduism combined science with spirituality and produced the world's greatest scientists. Between 3000 and 2000 years ago, mathematician Baudhayana calculated the value of pi, worked out theorems and the square root; Sushruta was a plastic surgeon and ophthalmologist; Charaka was the father of medicine; Aryabhatta, an astronomer and mathematician, said the world was round, invented the numerals and discovered zero; Kanada discovered gravity; and Varahamihira was a biologist and astronomer. There were many others too. These mathematicians and scientists lived in forests which were their source of inspiration. Their knowledge and enlightenment is still available, in beautiful poetry, for the seeker. They were also gurus who produced students to take forward their message of the equal importance of science and spirituality. Unfortunately, the limited number of students in the *gurukul* system and the lack of patronage in the medieval period prevented the dissemination of this knowledge, which gradually disappeared into oblivion.

Every aspect of nature is sacred for the Indic religions: forests and groves, gardens, rivers and other waterbodies, plants and seeds, animals, mountains and pilgrimage centres. The sacred is still visible in modern India in certain aspects of people's lives and in rural areas, especially among communities like the Bishnois, and many tribes, and in the many festivals which celebrate nature and the environment in so many ways.

Tamil Sangam literature belonging to the period between 300 BCE and 300 CE describes a scenario where people were seen as one of the components of five different ecosystems. Each ecosystem had its own unique habits of hunting, gathering, cultivating and worshipping deities. It appears that the ancient deities of Tamil Nadu continue to be worshipped in villages under different names. Although some of the deities may not be associated with an extensive forest cover any longer, most are found in intimate association with at least a small grove of plants or sacred groves.

Sangam literature describes the *aindu tinai*, the fivefold division of the geographical landscape. These landscapes are *kurinji* (mountains), presided over by Lord Murugan or Karttikeya; *mullai* (forests), whose reigning deity was Lord Krishna; *marutham* (agricultural lands), ruled by Lord Indra; *neithal* (coastal regions), the world of Lord Varuna; and *paalai* (wasteland/desert), which was the region of Goddess Kotravai (Durga). Each *tinai* has its own characteristic flowers, trees, animals, birds, climate and other geographical features. Of these, trees have played an important role in the social, cultural and religious aspects of ancient anthologies. Flowers are associated with gods and goddesses and the tradition of offering it to them finds mention in Sangam literature.[11]

The tradition of the sacred environment lives on to this day, from the sacred forests which may not be destroyed, to the sacred animals which may not be killed, to the sacred rivers which give life and the sacred mountains that inspire awe and reverence.

The river was the source of life and the earliest civilizations were found along its banks. The Ganga is the holiest of rivers and a dip in the river is believed to wash away one's sins. She is worshipped as a goddess, with the crocodile as her vehicle. She comes cascading down the hills from Lord Shiva's top knot. Even her presence on the earth is a divine response to the penance of King Bhagiratha.

But long before Ganga attained that status, the Sapta Sindhava (seven rivers) (*Rig Veda*, II.12; IV.28; VIII.24) played an important part in Vedic hymns. They are located in north India/Pakistan. They include the Sarasvati, Sindhu and the five major tributaries of the Sindhu—Shutudri, Parushni, Ashkini, Vitasta and Vipasa. The Sapta Sindhava were bounded by the Sarasvati in the east, Sindhu in the west and the other five in between. The 'Nadistuti Sukta' of the *Rig Veda* (X.75) contains a geographically ordered list of rivers, beginning with the Sarasvati in the east. River Sarasvati is now identified with the Ghaggar-Hakra.[12] Even after the river disappeared, probably due to seismic activity in the region, she continued to remain sacred and is remembered as the Goddess Sarasvati of learning and wisdom. In later literature, all the rivers of India attained sacred status.

The mountains are the homes of the gods. These awe-inspiring rocks have magical qualities that set them apart, such as the Kailas, the home of Shiva, in Tibet; the Vindhyas, which stopped growing at the request of Agastya, whose return they still await; Sheshachala, the home of Lord Balaji-Venkateshwara, and many others that dot the Indian countryside.

These are but a few examples of the sacred in nature. Animals, flowers, rivers, seeds and even entire cities are sacred, given their intimate association with nature.

Heritage may be defined as the cultural, social and spiritual values of our ancient past which still have relevance and which we must bestow on the coming generations. Indian culture is noted for its deep respect for Mother Nature. Even today, ancient traditions, customs and practices continue to flourish in our day-to-day life. Hinduism is noted for its deep respect for all forms of nature and the unique role that each life form plays in the ecology of the earth. The celestial River Ganga, the pristine Himalayas, the imposing Mahabodhi tree at Bodhgaya, the sacred blackbucks of the Bishnois of the Rajasthan desert, the *ki law kyntang* (sacred forests) of Meghalaya and the numerous sacred forests and gardens, rivers and kunds, animals and plants are all examples of our rich ecological heritage.

Religious practices are influenced by local environmental factors which play an important role in determining the relationship between man and god. Religions that consider that nature was created for the benefit of man and that man is the master of nature permit ruthless exploitation of natural resources. Primitive and nature-sensitive cultures always have some tenet in their religion about the sanctity of natural wealth. All religions have their key festivals coinciding with natural phenomena. Most studies of the ecological aspects of nature did not leave a record in stone but in the language of all in the oral tradition. They refer to the five traditional elements of eastern and western traditions: earth, water, air, fire and ether. These are represented by the five-headed snake which is common in Hindu and Buddhist cultures.[13]

Dr Karan Singh observes, 'All creation, whether this tiny speck of cosmic dust that we call our world or the billions of galaxies that stretch endlessly into the chasms of time, is in the ultimate analysis a manifestation of the same divine power. Thus, despite the multitudinous manifestations of this space-time continuum, there is ultimately no dichotomy between the human and the divine.'[14]

In 2002, when C.P.R. Environmental Education Centre (CPREEC) started a website on 'Conservation of Ecological Heritage and Sacred Sites of India', it also took up the documentation and publication of the ecological traditions of India. Till now, thirteen states have been covered. Having conceived and been closely associated with both these projects since their inception, I have collected a wealth of information, some of which I have used in this book, to illustrate the close symbiosis between Hinduism and nature, which has made India unique.

> *Sarvepi sukhinah santu, Sarvesantu niramaya,*
> *Sarve bhadrani pashyantu, Ma kaschit dukhamapnuyat*
> (May all be happy; May all be free from fear;
> May all see only good; May no one be unhappy.)
>
> (*Brihadaranyaka Upanishad*, 1.4.14)

This concept of welfare of all creation has been beautifully described by Karan Singh:

> Welfare is described not in limited terms but as all-embracing, covering not only the human race but also what, in our arrogance, we call 'lower' beings—animals and birds, insects and plants, as well as 'natural' formations, such as mountains and oceans. In addition to the horrors that mankind has perpetrated upon its own members, we have also indulged in a rapacious and ruthless exploitation of the natural environment. Thousands of species have become extinct, millions of acres of forest and other natural habitat laid waste, the land and the air poisoned, the great oceans themselves, the earliest reservoirs of life, polluted beyond belief. And all this has happened because of a limited concept of welfare, an inability to grasp the essential unity of all things, a stubborn refusal to accept the earth not as a material object to be manipulated at will but as a shining, spiritual entity that has over billions of years nurtured consciousness up from the slime of the primeval ocean to where we are today. Such an interpretation of welfare applicable to the whole of creation is perfectly consistent with the Vedic view of the unity and interrelatedness of all that is—God, self, the natural world—and although the

ancients may not have been as romantically enamoured of nature as they are sometimes made out to be, they undoubtedly believed themselves to be much more a part of it than their present-day successors.[15]

Hinduism has a cosmic, rather than anthropocentric, view of the world, an ontology sharply different from the Abrahamic religions which believe that 'God created mankind in his own image' (Genesis, 1:27) and 'let them have dominion over the fish of the sea and over the birds of the heavens and over the livestock and over all the earth and over every creeping thing that creeps on the earth' (Genesis, 1:26). Hindu traditions acknowledge that all life forms—humans, animals and plants—are equal and sacred, and thus appropriately placed to take on contemporary concerns like deforestation, intensive farming of animals, global warming and climate change.

Our country lies on top of the Indian tectonic plate. Its defining geological processes commenced seventy-five million years ago, when it was a part of Gondwana, the super continent, and began drifting in a north-easterly direction across the Indian Ocean. Subsequently, the subcontinent collided with the Eurasian plate, giving rise to the Himalayas, the world's highest mountain ranges, which are still growing. In the seabed immediately south of the Himalayas, the plate movement created a vast trough, which was filled with sediments borne by the Indus, Ganga and their tributaries, forming the Indo-Gangetic Plain.[16] The movement of the subcontinent towards the Eurasian plate has resulted in vast tectonic events, resulting in the appearance and disappearance of rivers—the appearance of the Indus-Sarasvati, Ganga and their tributaries, the disappearance of the Sarasvati, and her tributary Yamuna turning towards the Ganga. Hinduism, in its earliest forms, developed on the banks of these rivers, so the history of nature and the religion is closely intertwined.

The Allchins summarize the main reasons for the rise and decline of the Indus civilization as follows:

> The whole question of interpretation of the South Asian evidence is complicated by the rapid uplift of the Himalayas, Tibetan plateau, Pamirs and other mountain ranges . . . The rapidity, scale and recent date of this activity has only become apparent since the role of plate tectonics or 'Continental Drift' as a major factor in the shaping of these regions has been recognized. The relationship of past climates to such major tectonic activity is highly complex. At present all that can be said is that these changes must have had profound and far reaching effects not only upon the mountain regions but upon the whole subcontinent.[17]

In recent times, the deteriorating environment and the rise in activities threatening the ecological balance has revived an interest in environmental studies, and the relation between human activities and environmental degradation. Unfortunately, unchecked deforestation has led to severe natural disasters. The pollution of water sources—rivers and lakes—has limited the availability of clean water for drinking and irrigation. Floods and droughts are on the rise. Wildlife is under acute threat.

The World Bank Group commissioned the Potsdam Institute for Climate Impact Research and Climate Analytics to look at the likely impacts of temperature increases from 2–4° Celsius in South Asia. The scientists used the available evidence and advanced computer simulations to arrive at the results. Some of their findings include a warming climate, decline in monsoon rainfall and increased frequency of heavy rainfall events.[18]

Many places have become drier since the 1970s with increased frequency of droughts, leading to farmers' suicides. Heat waves have resulted in a substantial rise in the mortality of people and animals. As over 60 per cent of India's agricultural lands are rain-fed, the country is highly dependent on groundwater, which is already overexploited.

Most Himalayan glaciers, which feed the northern rivers, have been retreating. The Indus and the Ganga-Brahmaputra-Meghna basins are major trans-boundary river basins, and increasing demand for water is leading to tensions among countries over water sharing.

Climate-related impacts on water resources can undermine the two dominant forms of power generation in India— hydropower and thermal power—both of which depend on adequate water supplies to function effectively. Many parts of India are already experiencing water stress.

Coastal flooding and unplanned urbanization have increased the risk of seawater intrusion. Several islands in the Sundarbans have been submerged already due to rising water levels.

Malaria and other vector-borne diseases, along with diarrhoeal infections, a major cause of child mortality, are likely to spread into areas where colder temperatures had previously limited transmission.

It is unfortunate that it has become necessary to enact laws to control environmental degradation. The Environment (Protection) Act of 1986, Air Act, Water Act, Forest Acts, Wildlife (Protection) Act, Prevention of Cruelty to Animals Act, Biological Diversity Act, and many more are among the efforts of the state to protect the environment. But why has this become necessary? We have protected the environment over several millennia. This was done by respecting natural resources as a gift of the divine. When that veneration goes, the state is forced to act as a policeman.

Faith groups can play an important role in the effort to protect nature and the environment. Hinduism, with its ancient tradition of respecting nature, should re-invoke its rich heritage to ensure that people, animals and all of nature live together in harmony, and recreate the beautiful environment of ancient India.

Today there is a renewed regard for the religious beliefs which protect the environment. The Madras High Court recently said that opposition to worshipping the *panchabhuta* (five elements of nature) in the guise of promoting rationality is a reason for environmental degradation. Justice S. Vaidyanathan said that religious beliefs are protective of human civilization and the environment. 'Our tradition and values, passed down to us from our ancestors, are not wrong beliefs. They are scientific, rational and logical. That is why they worshipped nature. Even now, many of them who follow our ancestral beliefs continue to do so as they have got abundant sanctity.' Referring to people worshipping soil, fire, water, space and air, the learned judge said: 'It is not at all irrational. When nature gets the sanctity, it will not be ruined . . . Thus, nature was protected in those days. However, in the name of rationality, religious taboos were violated, the result of which we suffer these days.' These observations were made while disposing of a writ petition filed by the owner of a commercial building in Chennai.[19]

2

Groves and Gardens

(In that sacred forest)
All the animals and birds gave up their hostility.
The trees produced (flowers and fruit in such abundance)
That visitors received whatever they desired.
In the new cottages made of leaves,
The sacred fire burned all the time,
Making holier the holy forest.

—Kalidasa, *Kumarasambhavam* (5.17)

Forests have always been central to Indian civilization, representing the feminine principle in prakriti. They are the primary source of life and fertility, a refuge for the wanderer and a home for the seeker, and have always been viewed as a model for societal and civilizational evolution.

Forests were places of retreat, a source of inspiration, for all Vedic literature was revealed to the sages here. Rama's entire journey from Ayodhya to Lanka was through forests. In the Mahabharata, the big war is for urbanization and to capture the cities of Mathura, Hastinapur and Indraprastha. Yet the Pandavas spent their years of exile in the forest and made marriage alliances with forest tribes, a move that would help them later in the Kurukshetra war. They also learnt several important lessons from living in the forest, which became a source of knowledge and a place for learning higher truths. There were several classifications of the forest. The ancient forests have survived as the sacred groves of modern India.

The seals of the Indus civilization contain figures of wild animals such as the elephant, water buffalo, rhinoceros, deer, gazelle, antelope, wild sheep and goat and ibex and tiger,[1] which means that the area was once covered with dense forests. Rhino habitat ranges from open savannah to dense forest, while tigers live in swamps, grasslands and among trees, bushes and tall grass which camouflage them. Elephants are found in savannah and forests, where they can find fresh water to cool their thick dark skins. The large number of such seals suggests that the Indus–Sarasvati region was once a thick forest, not the agricultural fields or deserts we see today.

The Vedas were composed in the Indus–Sarasvati region. In these texts, there is a fundamental sense of harmony with nature, which, in turn, nurtured a civilizational value. Forests were the primary source of life and inspiration, not a wilderness to be feared or conquered.

The Vedas were written by sages living in the forest who saw it as a home and a source of revelation, exaltation and creativity. Some of the greatest verses of philosophy were written in forests. People drew intellectual, emotional and spiritual sustenance from the twin concepts of srishti and prakriti.

'So may the mountains, the waters, the liberal (wives of the gods), the plants, also heaven and earth, consentient with the Forest Lord (Vanaspati) and both the heaven and earth preserve for us those riches' (*Rig Veda*, VII.34.23).

Aranya means forest. The early Vedic literature includes the Aranyakas, which represent the earlier sections of the

Vedas, the speculations of the philosophy behind rituals. Aranyaka may be defined as 'produced by or relating to the forest' or 'belonging to the forest'. They were composed by sages living in the forest.

One of the most beautiful hymns of the *Rig Veda* is dedicated to Aranyani, the goddess of the forest. She is an elusive spirit, fond of solitude, and fearless. The poet asks her to explain how she can wander so far from civilization without fear or loneliness. He creates a beautiful image of the village at sunset, with the sounds of the grasshopper and the cicada and the cowherd calling his cattle. She is a mysterious sprite, never seen, but her presence is felt by the tinkling of her anklets and her generosity in feeding both man and animal:

> Aranyani Aranyani, who are, as it were, perishing there, why
>   do you not ask of the village? Does not fear assail you?
> When the chichchika (bird) replies to the crying grasshopper,
>   Aranyani is exalted, resonant, as with cymbals.
> It is as if cows were grazing, and it looks like a dwelling, and
>   Aranyani, at eventide, as it were, dismissed the wagons.
> This man calls his cow, another cuts down the timber,
>   tarrying in the forest at eventide, one thinks there is a cry.
> But Aranyani injures no one unless some other assails;
>   feeding upon the sweet fruit, she penetrates at will.
> I praise the musk-scented, fragrant, fertile, uncultivated
>   Aranyani, the mother of wild animals
>
> (*Rig Veda*, X.146. 1–6)

Aranyani never returns in later Sanskrit literature or modern Hinduism, yet her spirit pervades the goddesses of Hinduism: Prakriti, or nature; Bhu, the earth goddess; Annapurna, the giver of food; and Vana Durga, the goddess of the forest. In Bengal, she is worshipped as Bonbibi, the lady of the forest; in

Comilla, Bangladesh, as Bamini; in Assam as Rupeshwari; in Tamil Nadu as Amman; and so on.

Vanaspati, the lord of the forest, is invoked to taste the ritual offering and take it to the gods (*Rig Veda*, X.70.10). He is asked to sweeten the sacrificial oblation with honey and butter. He is the protector: 'May Vanaspati never desert us nor do us harm: may we travel prosperously home until the stopping (of the car) until the unharnessing (of the steeds)' (*Rig Veda*, III.53.20). The tree is the 'Lord of the Forest (Vanaspati), the shedder of nectar, and rejoicing the generations of men (is present) in the midst of our sacred rites' (*Rig Veda*, IX.12.7).

The *Rig Veda* says that plants are 'those that grew in old times . . . much earlier than even the devas . . . and are different from many different places' (X.97). There is knowledge that plants have life. 'As (the tree) suffers pain from the axe, as the shimal flower is cut off . . . so may my enemy perish' (III.53.22), and Indra, hero of the *Rig Veda*, lives in the forest: 'He (Indra) is in the forest . . .' (I.55.4).

In the Vedic period, trees were compared to gods and humans. The 'Aushadhi Sukta' of the *Rig Veda* addresses plants and vegetables as 'O Mother! Hundreds are your birth places and thousands are your shoots' (X.97.2). The *Atharva Veda* mentions the names of some herbs with their values. A medicinal herb is a goddess born on earth (VI.136). Later, this information became an important source of material for Ayurveda. The *Rig Veda* very specifically says that forests should not be destroyed (VIII.1.13). The *Atharva Veda* says: 'The earth is the keeper of creation, container of forests, trees and herbs' (XII.1.57–61). Plants are live (XII.1.57–61); 'Plants and herbs destroy poisons (pollutants)' (VIII.7.10); 'Purity of atmosphere checks poisoning (pollution)' (VIII.2.25); and 'Plants possess the qualities of all duties and they are saviours of humanity' (VIII.7.4).

The necessity for maintaining the ecological balance is clear in the Vedas. A verse from *Rig Veda* says, 'Thousands and hundreds of years if you want to enjoy the fruits and happiness of life, then take up systematic planting of trees.' These verses emphasize the importance of afforestation for survival, or else the ecological balance of the earth would be jeopardized. *Rig Veda* has dwelt upon various components of the ecosystem and their importance. 'Rivers occasion widespread destruction if their banks are damaged or destroyed and therefore trees standing on the banks should not be cut off or uprooted.'[2] The *Yajur Veda* also speaks of the ill effects of deforestation, while the *Atharva Veda* (V.28.5) says, 'The earth provides surface for vegetation which controls the heat buildup. Herbs and plants having union with sun rays provide a congenial atmosphere for life to survive.' The *Brihadaranyaka Upanishad* (3.9.28) equates trees with human beings: 'Just like a tree, the prince of the forest . . .'

The 'Bhoomi Sukta' of the *Atharva Veda* celebrates the role of the earth as the home of 'hills, snowy mountains, forests . . . who bears many plants and medicinal herbs . . . What forest animals of yours, wild beasts in the woods, lions, tigers go about man-eating—the jackal, the wolf, O earth . . .' (XII.1). The mention of both lions and tigers is significant because it suggests the presence of grassy plains, savannah, open woodlands and scrub country for the lion and subtropical forests for tigers.* The 'Bhoomi Sukta' is the best commitment to the environment in ancient Indian literature.

There were three categories of forests in the Vedas: *tapovana*, *mahavana* and *shreevana*.[3] The tapovana was a refuge

---

* The rock paintings of Bhimbetka, 10,000 years old, in Madhya Pradesh, contain both the tiger and the lion.

for meditation, an *abhayaranya* or sanctuary, where kings and commoners sought the guidance of sages. The ancient Indian civilization was nurtured in the forests which were the abodes of great rishis. The mahavana was the great forest in which all species could find refuge. The shreevana was the forest which provided prosperity. It was maintained by the temple and set aside exclusively for the practice of religion. These must have been the sacred groves of yore.

Parashara's *Vriksha Ayurveda* covers the basic role of plants. The chapter on 'Bijotpatti Kanda' mentions *vana varga sutriyani*, which deals with forest regions. Forests are described as *atavi, vipina, gahana, kanana, vana* and *maha aranya*. Parashara describes forests where trees, shrubs, creepers and grass grew naturally. *Caitraratha vana* was a beautiful sylvan tract frequented by *deva*s and *gandharva*s. Forests are classified according to their vegetation.[4]

As we move east, we learn more about the forest, especially from the Ramayana. The forest is made up of four sentiments: *shanta* (calm), *madhura* (sweet), *raudra* (angry) and *vibhatsa* (fearful). There are two forest types: the mahavana or larger forest represented by Chitrakuta and Dandakaranya, and the sub-forest of peace or tapovana represented by Panchavati. The forests are deciduous, and water occupies an important space.[5] The tapovana was the sacred grove where rishis lived in their ashramas.

Rama stays in four different forests during his exile. In the first phase, he stays at Chitrakuta, a deciduous forest filled with fruit-bearing trees like the mango and jackfruit. The next is Dandakaranya, situated in modern Madhya Pradesh, Odisha and Andhra Pradesh, also a deciduous forest with the sal, *badari* and *bilva* trees, among others. Dandakaranya is named after the *danda-trina* grass and is described as abounding in tall trees, sacred trees and secret fruit-bearing trees. Panchavati, where Sita was abducted, is a tropical dry deciduous forest, named after the *pancha* (five) *vata* or banyan trees. Kishkinda, where he visits the Pampa Sarovar or the lake situated between Rishyamukha Hill and Matanga Hill, is a dry and moist deciduous forest. Beyond these forested hills lay another vana, where the climate is very pleasant. The epic describes Kishkinda as a thickly forested area, quite unlike the barren region we see today. The last forest mentioned is situated in the trans-Himalayan region with alpine plants, the Aushadhi mountain of the Kailas, Rishabha and Mahodaya. The Dronagiri Hills, which has medicinal plants that Hanuman took back to Lanka, were situated between the Rishabha and Kailas. Finally, the author describes the evergreen forests of Lanka, situated off the Indian mainland.[6] It is amazing how much Valmiki knew and how little has changed in the forests that are found in each of these places, except for its density and wildlife. Rama also visited several ashramas in these forests, some of which may have formed the nucleus of future settlements.

The Ramayana is a botanist's delight, with detailed descriptions of forest types and the plants that grew therein. The forests were the home of many rakshasas or demons, probably a name for unfamiliar tribes. Rakshasas, asuras, *daityas* and *danavas* were famous tribes with great rulers, who once lived in these forests. Conflicts over the use of the forest arose between Rama and many rakshasas.[7] But the two gradually integrated. Ravana was the rakshasa king of Lanka; Kamsa, the king of Mathura, was the uncle of Krishna, a Yadava; Yayati, an ancestor of the Yadavas, married Sharmishtha, daughter of Vrishaparvan, a danava king; Bhima's wife, Hidimbaa, was a rakshasi; and so on. The word rakshasa comes from the root *raksha*, which means to guard, protect, save, take care of and preserve. They are enemies of devas in Vedic and epic literature. Among their many sins of commission, they disrupt the sacrifice (*Rig Veda,* VII.104.18), and several deities are invoked to kill them. There is a story that when Brahma created the waters, he created rakshasas to guard them. It appears that they were the inhabitants and protectors of the forest, who opposed the expansion of settlements that were destroying the forests.[8]

The most famous forest—then and now—was Naimisha Aranya. The Mahabharata was narrated to Shaunaka and other rishis by Ugrashrava Shrauti in the forest of Naimisha, which lay on the banks of the River Gomti between the kingdoms of Panchala and Kosala (near Lucknow in modern Uttar Pradesh). Rishi Shaunaka, it is believed, chanted all the verses of the epic in one breath. The Naimisha forest has several claims to fame. Lord Vishnu killed the demon Durjaya and his followers in this forest. When he threw his weapon, chakra, at the demon Gayasura, the demon's body was cut into three: one part fell at Naabhi Gaya in Naimisha, and the spot where the chakra or *nemi* fell is the chakra teertha or holy water (for bathing). Sage Dadhichi gifted his bones to devas to make the weapon vajra at Naimisha. Rama performed the *ashvamedha* ceremony in this forest, where Lava and Kusha chanted the epic Ramayana to their father and Sita vanished into the earth after uniting her sons with Rama. *Srimad Bhagavata Mahapurana* was recited here and Shri Satyanarayana *vrata katha* originated here. While the Pandavas visited this forest during their exile, Balarama visited it during his pilgrimage. Tulsidas composed his *Ramcharitmanas* here.

Naimisha Aranya appears to have been a magical place. In another version, Lord Brahma sent his *manomaya chakra* to the earth to show the sages the best place for meditation.

The chakra or nemi fell at chakra teertha, and the surrounding forest became Nemi or Naimisha Aranya. It is the most sacred of all teerthas.

Agni was the destroyer of forests. 'Agni, who of old thou burnt up Jaratha (a demon/tribe?) . . .' (*Rig Veda*, I.1.7); 'Agni . . . consumes the forest trees' (*Rig Veda*, I.140.2); 'The great devourer of plants' (*Rig Veda*, I.163.7) and so on. The forests could have been destroyed by deliberate or accidental burning. This was a tribal society that had learnt the use of fire and used it both for their rituals as well as for clearing the forest.

The best example of the clearing of forestland by burning can be seen in the story of the Khandava *dahana* or the burning of the Khandava forest. Khandava vana was an ancient forest situated west of the River Yamuna in modern Delhi, and inhabited by Takshaka and his tribe of nagas. Arjuna, with the help of Krishna, cleared this forest by setting it on fire to construct the new capital city of the Pandavas, Indraprastha, named after Indra, who lived here. The epic describes the terrible massacre of animals and birds as Khandava was burnt down:

And while the forest was burning, hundreds and thousands of living creatures, uttering frightful yells, began to run about in all directions. Some had particular limbs burnt, some were scorched with excessive heat, and some came out, and some ran about from fear. And some clasping their children and some their parents and brothers, died calmly without, from excess of affection, being able to abandon these that were dear to them . . . The tanks and ponds within that forest, heated by the fire around, began to boil; the fishes and the tortoises in them were all seen to perish. During that great slaughter of living creatures in that forest, the burning bodies of various animals looked as if fire itself had assumed many forms . . . And there came also, desirous

of battle, innumerable Asuras with Gandharvas and Yakshas
and Rakshasas and Nagas sending forth terrific yells. Armed
with machines vomiting from their throats iron balls and
bullets, and catapults for propelling huge stones, and rockets,
they approached to strike Krishna and Partha, their energy
and strength increased by wrath. But though they rained a
perfect shower of weapons, Vibhatsu struck off their heads
with his own sharp arrows . . . (Mahabharata, I.50).

There is no doubt where the writer's sympathies lay: with
the animals and people who perished in the conflagration.
This story illustrates the anger of local people towards the
destruction of their forest and the usurping of their land by
the Kuru prince. This was a man-made environmental havoc
for the sole purpose of building a new capital for the Pandavas.
It resulted in the migration of the remaining people, led by
Takshaka, out of the Khandava forest and the founding of the
city of Takshashila (modern Taxila). The conflict between
forest dwellers who wish to retain their sylvan surroundings
and politicians who want to acquire them for 'developmental
projects', especially mining, continues even today in places
like Jharkhand, Chhattisgarh and Odisha.

Urban settlements were created by burning down forests.
Although Arjuna and Krishna revel in the destruction of the
Khandava forest, the Pandavas treat the other forests that they
visit during their years of exile with great respect, forming
alliances that would stand them in good stead during the great
war of Kurukshetra. In the Mahabharata, the Pandavas have to
traverse many forests during their exile. Bhima, for example,
marries the rakshasi Hidimbaa. Their son, Ghatotkacha, takes
part and is killed in the Kurukshetra war.

However, Vrajbhoomi, the region around Vrindavan,
always had a good environmental balance, achieved through

the relationship between human settlements, forests and water resources. Between the villages there would be three types of forest: sanctuaries, dense woodlands and sacred groves. The first type was called raksha, a sanctuary for wildlife where no human would enter. This type of forest corresponds to the mahavana. Dense forest was called shreevana. While the natural arrangement of trees and forests would not necessarily be disturbed, people could go there to collect dry wood, leaves, forest produce and a limited amount of green timber. As care for this woodland was the responsibility of the village communities and as their livelihood depended on it, they naturally conserved it. Finally came the grove, called *vana khandi*, maintained by the villages as places for religious observance, festivals and recreation. Recreation, such as singing or dancing or *jhulan* (swinging on a seat suspended from the branch of a tree), was associated with festivals like *rasaleela*, commemorating Krishna's dance with the cowherdesses on the banks of the River Yamuna.[9]

Forests were never far away from habitations, and the Chinese Buddhist monk Hsuan Tsang writes, in the seventh century CE, of forests near Kapilavastu and Kushinagara in north Bihar.[10]

It was perhaps to offset the destruction of forests that the Puranas exhort the reader to plant trees. The *Varaha Purana* says, 'One who plants a pipal (*Ficus religiosa*), a neem (*Azadirachta indica*), a banyan (*Ficus benghalensis*), two pomegranates (*Punica granatum*), two orange (*Citrus reticulate*), five mango trees (*Mangifera indica*) and ten flowering plants or creepers will never go to hell' (172.39). The practice of *vana mahotsava* (tree plantation ceremony) is an ancient tradition. *Matsya Purana* speaks about it. *Agni Purana* says that the plantation of trees and creation of gardens lead to the eradication of sin. In the

Padma Purana (56.40–41) the cutting down of a green tree is an offence punishable in hell. Thus protecting the forest and expanding it was regarded as a sacred duty.

Buddhist traditions attribute Gautama's enlightenment to his meditation under the pipal tree, which was his tree of enlightenment. His first sermon was in the deer park at Sarnath, obviously a protected area. In the Buddhist period, patches of vegetation were preserved as *amara vana, venu vana, salai vana, ashoka vana, kadamba vana, vilva vana,* etc., named after the dominant tree species.[11]

People visited rishis and thinkers, including the Buddha, in the forest, for spiritual and moral guidance. Even Buddhist monks lived in monasteries in the forest, albeit along trade routes, as the Buddhist *chaitya*s and *vihara*s, built into caves along the Western Ghats, indicate. Indian philosophical discussions took place in the forest, rather than in villages, towns and cities.

While the Vedas hail the forest as a mystical and spiritual place of retreat, the Ramayana describes the types of forest, and the Mahabharata describes the Pandavas traversing and clearing forests. It is left to Kautilya's *Arthashastra* to describe the management of the forest.

In the Mauryan Empire of Chandragupta, forest was state property and its use had to be supervised by state officials, not unlike the forests of contemporary India, whose use is supervised by officers of the Indian Forest Service (IFS). There was a separate department headed by a head *samaharta*, like a collector, with several subordinate officers under a superintendent of the forest whose duty was to look after every matter related to the forest (II.6.1–2), collect and process forest produce like timber, fruits, fibres, medicine, etc., fix the price and sell it at the proper time (II.17.1–3). He was responsible for providing water and irrigation facilities to the forest during drought and other seasons. The superintendent of the forest could impose penalties on erring officers and antisocial persons misusing the forest produce or destroying the vegetation. The samaharta had to collect the revenue derived from forest produce and trade, and increase the economy of the state (II.6.28). The *akshatapala* was the controller of accounts and audit who maintained and checked the records of income and expenditure of the forest department (II.1.20). The entire system was integrated in such a way that each one was a check for another. The result was the luxurious growth of the forests all over the subcontinent during Kautilyan times. The forest management system, like everything else in the *Arthashastra*, had an excellent set of checks and balances.

The *Arthashastra* describes different forest types: *mriga vana* (forests of deer), *dravya vana* (economic forests), *hasti vana* (forests of elephants), *pakshi vana* (bird sanctuaries) and *pashu vana* and *vyala vana* (forests of wildlife), the last reserved for tigers and wild animals. The dravya vana was a source of forest produce, while the hasti vana was a sanctuary for elephants. Kautilya also mentions the *brahma aranya* (forest where the Brahmins could continue their studies of the Vedas and other scriptures); *soma aranya* (forest fit for carrying out religious sacrificial rites) and tapovana (forest of hermitages for ascetics). Deforestation and illicit tree felling were punished by *deya* (levy) and *atyaya* (fine). Ecological balance was maintained as there was environmental awareness. Protection of different species of animals was an important duty of the state (II.2.50).[12]

Thus the forest was recognized as a source of resources to be exploited during the reign of Chandragupta Maurya, when Kautilya (or Chanakya) lived. From Kautilya's description, the forest was also a buffer zone to dissuade neighbouring states from attacking.[13]

As a wilderness and a place of spiritual inspiration in ancient Vedic India, to the description of different types of forests in the Ramayana, to the management of it in Kautilya's *Arthashastra*, the forest has come a long way. It must have been burnt down for the construction of cities in Indus civilization, but they were generally protected.

An important aspect of nature worship was the protection of some patches of forest dedicated to local deities or ancestral spirits. These were the manifestation of the spiritual and ecological ethos of indigenous communities. The 'Prithvi Sukta' hymn of the *Atharva Veda* says: 'O Earth! Pleasant be thy hills, snow-clad mountains and forests; O numerous coloured, firm and protected Earth! On this earth I stand, undefeated, unslain and unhurt' (XII.1.11). Another hymn says: 'Whatever I dig out from you, O Earth! May that have quick regeneration again; may we not damage the vital habitat and heart' (XII.1.35).

In Kalidasa's *Abhijnana Shakuntalam*, there is a reference to the importance of forests and the preservation of wildlife, and the symbiotic relationship between people and the forests in the ashrama life. This was the tapovana, a peaceful and tranquil part of the forest where all forms of life could live in complete harmony. Kalidasa's evocative description of prancing deer and singing birds, of flowers in bloom and leafy creepers evokes an image of an idyllic forest rich in plant and animal life. This is the sacred grove of India's countryside and ecological tradition, a rich heritage of flora and fauna preserved out of faith and belief in a changing world. It represents local folklore and religion. Every village has a grove which is a protected area associated with local folk deities of uncertain origin.

Sacred Groves

Sacred groves are the home of local flora and fauna, a veritable gene pool of animal, insect, bird and plant species and a mini-biosphere reserve. The rich plant life helps to retain subsoil water and, during the hot summer months, the pond in the grove is often the only source of drinking water. The groves are a unique form of biodiversity conservation, and are living examples of the Indian tradition of conserving the ecology as a natural heritage.

They are an area of conservation as well as a spiritual retreat. They probably represent the singlemost important ecological heritage of the ancient culture of India. Sanskrit and Tamil literature are full of references to forests where wise and holy men lived, but the tradition probably goes back further in time, to food-gathering societies who venerated nature and the natural resources on which they depended for their existence. Sacred groves are the tapovana that once existed within the forests of ancient India, where the ashramas of the rishis were located. The tapovanas were inviolate, unlike the mahavana where man encountered and hunted powerful predators, or the shreevana, which were exploited for their resources. Other sacred groves go back to the time of pre-agricultural hunting and gathering societies, before human beings settled down to till the land or raise livestock. At the dawn of civilization, primitive men believed that deities resided in stones, trees, animals and woods, and thus this was their way of expressing gratitude to and respect for nature for providing them food and services. Images and temples came much later. Sacred groves are perhaps as old as civilization itself, born at a time when pristine religion was taking shape.[14] They exemplify the perceived interlink between man and his natural environment

as well as his ecological prudence,[15] provide a cultural identity
to each community, represent native vegetation in a natural or
near-natural state, and thereby contribute to biodiversity and
environmental conservation.[16]

Sacred groves have served as significant reservoirs of
biodiversity, conserving unique species of plants, insects and
animals. The sanctity attached to specific trees, mountains,
rivers, animals, caves and sites continues to play an important
role in the protection of local biodiversity. Plants are used
by tribal healers and priests who take a strong interest in the
preservation of such ecosystems. The belief that spirits inhabit
remote areas has served to quickly regenerate abandoned and
hidden plots into mature forests.[17]

Culture and environment have always been closely
interlinked. The myths and traditions that venerate plants and
animals, forests, rivers and mountains have played a key role
in protecting and preserving India's biological diversity over
centuries. The close connection between nature and divinity
has been an important part of the Indian religious ethos.

Sacred groves also unify the faith systems of India, uniting
the tribal and the philosopher, the cattle grazer and agriculturist.
A sacred grove may be a patch of trees left on the outskirts of
villages in the plains, or a part of a forested area, dedicated
to local folk deities or ancestral spirits.[18] They are protected
by local people through ancient traditions and taboos which
incorporate spiritual and ecological values. Thousands of such
groves have survived from ancient times, as repositories of
rare and varied local biological diversity, with an important
ecological role.

They are also cultural markers of the people and their
relationship with the environment. The existence of sacred
groves gives a fresh insight on environmental, historical and

sociocultural information. Woods, forests, rivers, streams, rocks, mountains, peaks and trees belonging to ancestral spirits or deities are found throughout the world.

The ENVIS Centre on the Ecological Heritage and Sacred Sites of India at CPR Environmental Education Centre has documented, till date, 10,377 sacred groves from across India: Andhra Pradesh, 677; Arunachal Pradesh, 159; Assam, 29; Bihar, 43; Chhattisgarh, 63; Goa, 93; Gujarat, 42; Haryana, 57; Himachal Pradesh, 329; Jammu and Kashmir, 92; Jharkhand, 29; Karnataka, 1476; Kerala, 1096; Madhya Pradesh, 170; Maharashtra, 2820; Manipur, 166; Meghalaya, 105; Odisha, 188; Puducherry, 108; Rajasthan, 560; Sikkim, 16; Tamil Nadu, 1275; Telangana, 57; Uttarakhand, 133; Uttar Pradesh, 32; and West Bengal, 562.[19] Another estimate suggests that the number of groves in the country may be as high as 1,00,000 to 1,50,000.[20]

Generally, each grove is attached to a village, a community or a tribe which preserves it as a repository of native plant, animal, insect and microorganism species. This is a testimony to the efforts of the local communities to protect their natural forests from clearing for the purpose of agriculture and settlements.

Sacred groves have been reported from various regions and ecosystems of India. A diverse range of ecosystems is preserved as part of this tradition, along with regional and local identities represented in the names, practices and management of the groves.

Travelling from north to south, the sacred groves of Jammu and Kashmir are managed by religious bodies or management committees. *Bani* means forest and *dev bani* means sacred forest or grove. Baba Roachi Ram, Bua Sjawati-ji, Bua Dati-ji, Lord Hanuman, Mata Vaishno Devi, Peer Baba and Raja Mandlik-ji are some of the deities to whom these groves are dedicated.

In larger groves, normal forestry operations are carried out and the income goes to the shrine. Small groves are highly protected and to remove anything from it is a taboo. People living around them voluntarily protect them. However, one wonders at the state of these groves after the migration of the Pandits and the growth of terrorism in the state.

In Haryana, unlike other states, there is no generic name for sacred groves, although the sites are protected for similar reasons. Khetanath, Jairamdas, Shiv, Bala Sundari, Nao Gaja and Mani Goga Peer are some of the deities to whom these sacred forests are dedicated. They act as a repository for medicinal plants and as a source of fruits and water.

In Himachal Pradesh, the local myths and legends associated with sacred groves preserve them. There are several *dev van* or *devta ka jungle*, sacred groves where trees may not be cut or dry leaves carried outside the grove. Bakhu Nag Devta, Ringarishi Devta (named after an ancient sage) and Devi are the deities to whom these groves are dedicated. The thick forests provide a good habitat for wildlife. There are about 10,000 temples in the state with well-defined management committees and *biradari* panchayats (caste councils). The important deities of Himachal have groves attached to their temples.

The sacred groves in Uttarakhand are locally known as *deo bhoomi* and *bugyal,* a high-altitude alpine grassland or meadow. The groves may be dedicated to Chandrabadni Devi, Hariyali Devi, Kotgadi Ki Kokila Mata, Pravasi Pavasu Devata, Devrada or Saimyar. They are a gene pool of indigenous species and a source of rich plant diversity. Rituals and traditional practices in these groves play a crucial role in fostering threatened species like the griffon vulture.

Sacred groves are known as dev van or just van in Uttar Pradesh. Samaythan, Vansatti Devi, Bhairav Baba, Phoomati

Mata, Shiv and Ram–Janaki are some of the deities to whom they are dedicated. These groves hold special significance in improving soil fertility through biomass build-up, nutrient cycling, conservation of moisture and providing a deep-root system with soil-binding properties. Probably the most famous are the sacred forests of Vraj, where Krishna grew up, playing with his friends and performing the rasaleela dance with the *gopis*. There are twelve forests left out of what was once several hundred. The most important is *vrinda vana*, the forest of *vrinda* or tulsi. Although they are all associated with the story of Lord Krishna, Vrinda Devi is the reigning goddess, installed in the *kamya vana*.

The sacred groves in Bihar are locally known as *sarna*s. The groves are dedicated to the deities Raksel, Darha, Marang Baru, Jaher Buri, Chandi, Dharti, Satbahini and Jahera. Hatubongako, the village gods of the groves, are regarded as the guardians of the village and their help is invoked in agricultural and other economic activities.

The tribals of Jharkhand worship their sacred groves, which are also known as sarnas. Such a grove must have at least five sal (*Shorea robusta*) trees, held to be very sacred by the tribals. Non-tribal Hindus also worship in such sarnas, which they call *mandar*. The Sarhul festival is celebrated in the *sarhul sarna* when the sal trees start flowering. The sacred groves celebrate the importance of this tree in their culture. Sarna means a grove, generally of sal trees which grow in abundance in Jharkhand. It is in these groves that the tribes of Chota Nagpur venerate their god and their spirits.[21]

In West Bengal sacred groves are known as *gram than, Hari than, Sabitri than, jaher than* (shortened to *jahera*), *santalburi than, shitala than, deo tasara* and *mawmund* (a grove or a group of trees). Sitala, Manasa, Devimani (lady of the grove) and

Ma Kali are the deities to whom these are dedicated. The sacred grove is associated with a range of oral narratives and belief systems. These make up a unique social means to prevent intra-group conflicts and violation of the grove by outsiders.

In Sikkim, *gompa* means monastery, which is managed by the gompa authority or lamas, or by the village community.[22] Sacred groves in Sikkim are attached to Buddhist monasteries and are called Gompa Forest Areas (GFAs). The groves are dedicated to local deities: Cho Chuba, Loki Sharia, Guru Padmasambhava and Rolu Devi Than. The highlands of Demojong below the Kanchenjunga peak is the most sacred site.

Forest dwelling tribes such as Bodo and Rabha inhabiting the plains and foothills of western Assam have the tradition of maintaining sacred groves which are locally called *than*. Dimasa tribes of the North Cachar Hills of Assam call them 'Madaico'.[23] Sibrai, Alu Raja, Naikhu Raja, Wa Raja, Ganiyang, Braiyung and Hamiadao are the various deities to whom these sacred groves are dedicated. Vaishnav temples called Sankaradeva Maths, to which sacred groves are attached, are distributed all over the state.

In Arunachal Pradesh, the gompas or sacred groves managed by lamas and the Mompa tribe are attached to the Buddhist monasteries and are called Gompa Forest Areas, like in Sikkim. They are dedicated to the local deities Ubro or Ubram and Thouw-gew. The monasteries with sacred groves are mainly located in the West Kameng and Tawang districts of the state. Fifty-eight GFAs were reported from these two districts and a few sacred groves from the Lower Subansiri and Siang districts. Ethnic groups have preserved and protected forest patches and even individual trees and animals due to their traditional respect for nature.

The worship and protection of forests called *umanglai*, because of their associated deities, is still practised by Manipuris who preserve their ancient tradition. Umanglai in Manipuri comes from two words: '*umang*' meaning forest and '*lai*' meaning gods and goddesses. These groves are locally known as *gamkhap* and *mauhak* (sacred bamboo reserves) and are dedicated to Umanglai, Ebudhou Pakhangba, Konthoujam Lairembi, Chabugbam and Chothe Thayai Pakhangba. Ecologically valuable species are found in several sacred groves of Manipur.

The local name for sacred groves in the Mizo language is *ngawpui*, meaning virgin or very old grove. Ka Niam is characterized as Ka Niam Tip Briew Tip Blei, that is, the religion of knowing man and knowing god. Man must behave well with his fellow men so that he may do god's will.[24]

The sacred groves in Meghalaya are known as *law lyngdoh*, *law niam* and *law kyntang* in the Khasi Hills, depending on the places where they are located. These are dedicated to Ryngkew, Basa and Labasa. Ancestral worship is traditionally performed here. *Khloo blai* in the Jaintia Hills and *asheng khosi* in the Garo Hills are owned by individuals, clans or communities and are under the direct control of the clan councils or local village authorities. *Law lum jingtep* is dedicated to ancestors and is the forest for graves.[25] In forested areas, the focus of worship is on ancient monolithic stones erected in memory of departed elders.

One of the most remarkable features of the Khasi Hills is its sacred forests, which have been preserved since ancient times. The most famous among them is the Mawphlang Sacred Forest, about 25 km from Shillong. It has various plants, flowering trees, orchids and butterflies and is an ideal destination for nature lovers. The groves which have been preserved since time immemorial are in sharp contrast to their surrounding grasslands. They are generally rimmed by a dense growth of *Castanopsis kurzii* trees, forming a protective hedge which halts the intrusion of Khasi pine (*Pinus kasia*) which dominates all areas outside the sacred groves. Inside the outer rim, the groves are virtually 'nature's own museum'. The heavily covered grounds have a thick cushion of humus accumulated over centuries. The trees in every sacred grove are heavily loaded with an epiphytic growth of aroids, pipers, ferns, fern allies and orchids. The humus-covered grounds likewise harbour myriad varieties of plant life, many of which are found nowhere else.[26]

Sacred groves are found all over western India. In Rajasthan they are called by various names such as *vani* in Mewar, *kenkri* in Ajmer, *oran* in Jodhpur, Bikaner and Jaisalmer and *shamlat deh* and dev bani in Alwar. The groves are dedicated to Garva-ji, Bharthari-ji, Naraini Mata, Peerbaba, Hanuman-ji and Naharshakti Mata. Rajasthan provides an ideal example of the support of the tradition of service to the ecosystem. The Gujjar people of Rajasthan have a unique practice of neem planting and worshipping it as the abode of god Devnarayan.

The word oran is derived from the Sanskrit word aranya which means forest or wilderness. Orans are sacred groves of trees set aside in Rajasthan for religious purposes. Considering the fact that the ancient River Sarasvati once flowed beneath the deserts of Rajasthan, and that Vedic literature came into existence on the banks of the Sarasvati, it is likely that the Aranyakas were composed in the aranya or orans of Rajasthan. In some orans, there are platforms covered with tiled roofs with *havan kund*s of different shapes, not unlike Vedic *homa kund*s. However, for the most part, oran today simply denotes common land with trees and some grass cover on it. In Rajasthan and Gujarat, the gods of the orans are known as Oranmata,

Devbani and Jogmaya. The orans in western Rajasthan, filled with *khejri* trees (*Prosopis spicigera*), deer, blackbuck and nilgai, are protected by the Bishnoi community, for whom they are sacred. In the year 1730, in the village of Khejarli in Jodhpur district, 363 Bishnoi women gave up their lives to protect the khejri trees, giving rise to the Chipko or 'Hug a Tree' movement.

Sacred groves are seen throughout Gujarat. Khodiyar Mata, Oran Mata, Jhalai Mata, Panch Krishna and Mahadev are some of the deities to whom these are dedicated. Sacred groves play an important role in the conservation of biodiversity, recharge of aquifers and soil conservation in this partially desert state. The cutting of trees and removal of wood are strictly prohibited.

Tribals form 19.9 per cent of the population of Madhya Pradesh in central India. Sacred groves, known as *deogudi* or sarna, conserve many plants and animals.[27] The groves are dedicated to the deities Bursung, Pat Khanda, Ganganamma, Mahadev, Bhandarin Mata and Danteshwari Mata. The tribals believe that if the groves are not maintained properly or are destroyed, natural calamities will ruin their clan.

In Chhattisgarh, sacred groves are locally known as *matagudi*, *devgudi* and *gaondevi*. Different tribes have their own Mata or Gaondevi (village goddess). Some of the deities to whom these groves are dedicated are Andhari Pat, Chala Pachao, Sarna Burhia, Sarna Mata, Mahadania and Budhadev. Sarna or jahera (sacred groves) are predominantly found in the Chota Nagpur region.[28] The community rituals are often at the time of the blossoming of the flowers of the trees in the groves and other agricultural operations, revealing the close harmony between nature and tribal communities. There have been constant clashes between tribals, many of whom have been

alleged to have joined the Maoists, and the state government.
The mahua, grown in the sarna, has been a major source of
income for the tribes, but its trade has been restricted, leading
to violent confrontations.

In Maharashtra, sacred groves are found in tribal as well
as non-tribal areas. The sacred groves in the western part are
called *devrai* or *devrahati*, whereas in the east, the Madiya tribals
call them devgudi. Some of the deities to whom these are
dedicated are Maruti (Hanuman), Vaghoba (tiger god), Vira
and Bhiroba (manifestations of Shiva), Khandoba (the patron
deity of several castes as well as the hunter-gatherer tribes who
live in the Western Ghats. He is regarded as a manifestation of
Shiva-Bhairava), Vetal (a ghost-like spirit), Mhasha and Shirkai
(local goddesses). Most of the cults associated with sacred
groves in Maharashtra are Mother Goddess cults—Kamaljai,
Mariai, Bhavani, Bhagvati and Tathawade.[29] Sontheimer traces
the origin of Khandoba to the worship of the anthill, the seat
of snakes, of Goddess Shirkai to Pune district.[30] The felling of
timber and the killing of animals in sacred groves is taboo.

In Goa, sacred groves are known by various names such as
devrai, *devran* and *pann* and are dedicated to the deities Durga
and Rashtroli. The tribes of Goa—Gavda, Kunbi, Velip and
Dhangar Gouli—worship various forms of nature. They still
maintain the tradition of revering sacred goats, sacred banyan
trees, sacred hills, sacred stones and sacred ponds, along with
the groves.

Pavitra vana, the local name of sacred groves in Andhra
Pradesh and Telangana, means a sacred forest. These are
small groves attached to each village that have been protected
over years by local people. Most of these are rich in flora
and fauna and have a good water source, like a spring, well,
waterfall, pond, river or stream. People have protected

them by attributing divinity and other supernatural qualities to them.[31]

Sacred groves in Odisha are recognized by names such as jahera and *thakurnam*. Jaheras are places of worship in Odisha located in forests outside the village, dedicated to the deities Moreiko and Turuiko (god of fire).[32] Other deities to whom these groves are dedicated are Jhakeri, Gram Siri, Gossa Pennu, Pitabaldi, Loha Pennu, Gaisri and Pat Baram.

The most notable community-conserved areas of Karnataka are its sacred groves. They vary in size, ownership patterns and vegetation. The groves come under two classes: smaller groves or *kan*s that are entirely protected and larger groves or *devarakadu/devarkan*, which function as resource forests, offering both sustenance and ecological security. The presiding deities to whom these groves are dedicated are usually Hulideva (tiger deity), Naga (snake), Jatakappa, Bhutappa and Choudamma, Mailara, Bhairava and Govardhan. A unique feature is the offering of terracotta hounds in the groves of Kodagu.

The sacred groves in Kerala are known locally as *kaavu* or *sarpa kaavu* (snake groves): there are *Ayyappan kaavu, Sastan kaavu* and *Bhagavati* or *Amman kaavu*, depending on the deities to whom these are dedicated. Serpent worship is an important feature of sacred groves in the state, as nearly all kaavus have images of snakes. Deities of kaavus managed by tribespeople are Yekshi and Vanadevata, the goddess of forests or spirits.

Kovil thoppu or *kovil kaadu* means 'forest of the temple' in Tamil. In Tamil Nadu, every village has a sacred grove ranging in size from 1 to 500 acres. They are known as kovil kaadu, *swami thopu, swami sholai,* Ayyappan kaavu (in Kanyakumari), *kaattu kovil* and *vanakkovil,* meaning forest temple (Figure

2.1). The deities associated with the groves are Aiyanaar,
Sastha, Muniswaran, Karuppuswami, Vedappar, Andavar and
goddesses like Selliyamman, Kali, Mari, Ellaikali, Ellaipidari,
Sapta Kannis, Pechiyamman, Rakkachiyamman and so on.
The groves are the repositories of medicinal plants. Many of
these are also important archaeological sites with evidences
of Palaeolithic or Neolithic cultures. Sittannavasal in
Pudukkottai district, for example, combines 3000-year-old
Neolithic dolmens, ancient sacred groves, an ancient water
tank, 1500-year-old residential caves of Jain monks and
1300-year-old painted Jain caves.

Figure 2.1: Poovanandi sacred grove at Kannoothu, Trichy

In the *Silappadikaram*, we learn that the Chola port city
of Poompuhar had a number of groves like *ilavandigai sholai*
(sacred grove), *kavera vanam* and *sampati vanam* named after
Sampati, the elder brother of Jatayu, and so on. Sampati vanam
is now known as Pullirukkuvelur or Vaitheeswaran Koil,
which was once a suburb of Kaveripattinam.

The Kollimalai Hills, situated in the Namakkal and Perambalur districts of Tamil Nadu, are closely linked with ancient Tamil literature. In the Tamil epics *Silappadikaram* and *Manimekalai*, there is an interesting reference to Kollipavai, the deity in the sacred grove, who is also the guardian of the forests. Apparently, the sages were looking for a peaceful place to undertake penance and chose Kollimalai. When they began their rituals, the demons invaded the hills to destroy their peace. The sages prayed to Kollipavai who, according to the myth, chased away the demons with her enchanting smile. She is still worshipped by the people here and her smile is revered. The Kollipavai Temple is located in one of the fifteen sacred groves here and can be approached only on foot.[33]

There is an ancient temple in the Kollimalai Hills dedicated to Lord Arapaleeshwarar on the River Aiyaru. Many sacred groves, guarded by local temple deities, are found in the forests near the Akashaganga Falls even today, where the felling of trees is strictly prohibited. The area surrounding the temple of Arapaleeshwarar and the sacred groves is still held in reverence. The groves where the temples were located are called *kaveri vanam* and *saya vanam*.[34]

Rajaraja I donated lands and paddy for the maintenance of the *Pidari* and *Ayyan kovil kaadugal* (sacred groves) (SII, Vol. II, p. 56). A festival called Vana Mahotsavam or Spring Festival was conducted on the full moon day in the sacred grove. The decorated idol was taken out in procession to the grove where it was kept for worship throughout the full moon night. The villagers were allowed to enter the grove and worship the deities on condition that they would not harm the animals and plants.

A festival is held once a year in the sacred groves of Tamil Nadu during which pongal, a mixture of rice, lentils and *vellam* (unrefined brown sugar), is prepared as *prasadam* (food which is a religious offering), cooked on dry twigs from the grove. Apart from this, twigs and branches of the groves cannot be plucked or used. It is mandatory that the grove is always kept clean: one is not allowed to urinate or defecate within the grove.

The groves were maintained by the priest(s) or village committees according to an inscription of the Chola king Kulothunga III, which records that a committee consisting of three representatives of the king along with the temple superintendent, the priest, organizer of festivals, the manager, the temple mason and the accountant were entrusted with the task of maintaining the temple.[35] An inscription of the sixth year of Kopperum-Singa-Deva records the gift of a grove, called Alagiya Pallavan Toppu, in Urrukkuruchchi in Kudal Naadu, by Alappirandan Alagiyasiyan Kopperum-Singan of Kudal in Kil-Amur Naadu, for supplying areca nuts, flower garlands, etc., to the god at Tirumudukunram in Paruvur-kurram, a subdivision of Irungolappadi in Merka Naadu, situated in Virudaraja-bhayankara Valanaadu. The garden of Alagiya Pallavan Toppu must have been named after the chief. Incidentally, Kopperunjinga II also bore this surname.[36]

Maravarman alias Tribhuvana Kulasekharadeva, in 1272 CE, gifted land for the lighting of a *nanda vilakku* and for food offerings to the goddess housed in the Varantaruvan Padaaran kaadu (forest of the boon-giving Padaran) of Pattivasekaranallur in Poliyurnadu.[37] Inscriptions belonging to Devaraya II dated 1424 CE and found on the south wall of the central shrine in the Masilamanishwarar Temple at Tirumullaivoyil, Saidapet taluk in the Chingleput district of Tamil Nadu, mentions that King Orri Mannan alias Udaiyar Orri-Arasar and Arasuperumal alias Kadavaraya (Devaraya II) gifted 4000 *kuli* of land for conducting certain special festivals in the Kadavarayar

Tiruttoppu (sacred grove).[38] The Vijayanagara kings organized festivals in *tiruttoppu* to establish their reverence for the groves and the deity within.

According to an inscription which can be dated to 1521 CE, Kandadai Madhavayyangar gifted 80 *pon* of gold, the interest from which was to be spent for the god Ranganatha when he halted at Madhavayyangar Tiruttoppu Mandapa, and on the fifth day of the Masi festival instituted in the name of Krishnadeva Maharaya, when the god halts at the garden adjoining *pradhani* Timmarasar Toppu.[39]

An inscription belonging to the period of Maravarman Sundara Pandyan I (1229–30 CE) refers to a gift of two pieces of land at Sikaranallur in Kunriyur Naadu as *tiruvettai tiruttoppu* to the temple of Tirunalakkunramudaiya Nayanar by Sankaran Kandan of Kulattur in Malaimandalam after purchasing the same from some local residents, in order to provide a garden where the image of the god would be taken in procession and worshipped on festive occasions. It records a further gift made by the same donor, by means of an agreement with the *naattaar* (village administration), for the supply of 10 *kalam* and 5 *kuruni* of paddy every year by the naattaar of Malai-nadu (probably by some investment with them), for the offering of *tiruppaavaadai amudu* to the deity after being taken in procession to the grove on the days of the festival in the months of Maargali, Maasi and Panguni.[40]

The sacred groves in Puducherry are locally known as *kovil kaadugal* and Ayyappan kaavu. Groves varying in size from 0.2 to 5 hectares have been documented in the state. Aiyanar, Poraiyatta Amman, Pachaivali Amman, Selli Amman, Kaliamman and Maduraiveeran are some of the deities to whom these are dedicated. The sacred groves are often dedicated to local spirits or deities and the people attach sanctity to them.

Religious practices and cultural traditions deter people from exploiting the biodiversity contained within them.

One of the important reasons for the continued survival of the sacred groves has been the taboos, beliefs and rituals along with folklore. In many places, women are forbidden to enter the grove during menstruation. In some groves, the practice of tonsuring the head and placing stone statues of the snake god is prevalent. Footwear, urination and defecation are all not permitted inside the grove as a matter of custom.

The deities protecting the sacred groves vary from state to state as we have seen. The deity may be male or female, animals or ancestors. At the dawn of religious thinking, deities were imagined by primitive societies to reside in stones, trees, animals and woods. They could be installed in a forest patch or even under a single tree.

Another common practice found in many states as far apart as West Bengal and Gujarat or Tamil Nadu and Bihar is the offering of terracotta horses of differing sizes to the deities. This is done in the belief that it will result in a good harvest. This tradition is suggestive of the ashvamedha. The horse is considered next in importance only to man, according to local people. The speed and strength of the animal is necessary for the protection of the village. It is gifted for the use of local spirits. For example, Aiyanaar in Tamil Nadu, it is believed, rides the terracotta horses around the boundaries of the village at night to safeguard the village. The terracotta horses of the state are huge, renowned for their beautiful decorations and colours (Figure 2.2). Perhaps the ritual of dedicating a horse to the gods of the grove dates back to the Rigvedic period and the ashvamedha, when the territory covered by the horse as it roamed for a year was claimed by the tribe. Although horses are never sacrificed, goats, fowl and buffalo are sacrificed in some groves, in fulfilment of a vow.

Photograph by M. Amirthalingam

Figure 2.2: Serakulathar sacred grove at Asur, Perambalur

The terracotta tradition is linked to Mother Earth being a symbol of fertility, and the many offerings to her are in fulfilment of vows for good health, a bountiful harvest and for the gift of life. The wealth of the grove, the richness of the plant life within and the life- and health-giving properties of the plants (which are generally medicinal) are all gifts of the earth, which is venerated as the Mother Goddess, the Great Earth Mother to whom the 'Bhoomi Sukta' of the *Atharva Veda* is dedicated. During the spring or summer season, week-long celebrations are conducted during which ritual prayers are offered to the deity. On these occasions, food is usually cooked using twigs collected from the grove.

Social and cultural mechanisms have played an important role in preserving the sacred groves. Over the centuries, the customs, rituals and ceremonies of the tribal communities have evolved and coalesced into their own unique culture. The conservation culture diversified into different forms of beliefs, rites, rituals, myths, taboos and folk tales. Festivals are generally organized by the whole community. Folk tales and folklore strengthen the cultural bond between the people and the grove. One of the main reasons why the local people do not plunder the sacred groves is the widespread belief among them that those who do so will invite the wrath of the deity. Taboos are an important means of social control in primitive societies. People do not harm the groves mainly for fear of the unknown, believing that those who cut a tree or use an axe there may be harmed by the presiding deity. Understandably, ancient cultures imposed restrictions and punitive actions primarily to stop the changing attitudes that would destroy the groves, which were preserved for various ecological reasons. The gods were invoked through rites, rituals and folk tales to create a fear of the consequences. The social divide is also reflected in how the preservation of certain groves is reserved for specific caste groups or tribes.

Resource extraction and alterations of land use are generally discouraged. Agricultural activities, erection of unauthorized structures and axing of trees are totally prohibited. However, in many groves, collection of fallen branches/trees and firewood for ceremonial cooking is permitted. Most people are not averse to using dead and fallen twigs for cooking within the groves during rituals and ceremonies. Carrying tools like swords, axes, knives, sickles, etc. is generally banned. The only metallic structures found within the groves are tridents, spears, swords, bells, etc., associated with the deity or the ritual. Non-vegetarian food can only be the meat of sacrificed animals, which can be served only on rare occasions when the animals are ceremonially offered to the deities.

The protection enjoyed by plants is extended to animals living within the grove, who may not be harmed, nor can

animals dedicated to the deity be harmed even when they stray into the village.

Rituals and festivals are an important component of the belief system, aimed at pleasing supernatural forces to ward off dangers such as drought, illness, epidemics, etc., and for seeking rich harvests and good health. Offerings are made ritually during festivals and these include offerings of terracotta horses, bulls and elephants, the last one being characteristic of the coastal groves of Tamil Nadu,[41] where elephants are not known to have existed.

Terracotta represents the powers of renewal inherent in the earth. It represents the Hindu philosophy of birth, death and rebirth. This is also the cyclic role of clay—it represents the votive offering for a certain period. As the clay slowly disintegrates and goes back to Mother Earth, it is time for the creation of a n? figure. The figurines can only be worshipped for a limited p? of time. In fact, a handful of clay is taken from the old fig? more clay added to it to make the new figure. A simila? is done during Ganesh Chaturthi, the festival of Ga? terracotta offerings of animals are always made of c? the open to go back to the mud they came from?

The relationship between human beings as civilization itself. Mud was used to cons? in, pots to store food and water and muc? Neolithic life depended on it. With t? new materials were used, but the ? continued till the present day, par? India. The sacred groves are in? potter and pottery. The clay animals are an inseparable pa?

Generally, the potter i? makes the terracotta figur?

living animals on behalf of his clients. His pots are broken at every birth and death, representing the renewal powers of the earth. In fact, a wedding cannot begin without the arrival of the pots. The potter performs the ritual of making the terracotta figures and the ritual of worship at the temple before the clay figures are offered to the deity.[42] His tools are few— the potter's wheel and his own hand. For the figurines, he uses a mixture of sand, husk and clay, unlike the mixture of sand and clay used for pots.[43] But the offering must be installed in a grove, in the open, under the skies.

There are important differences in the management of the sacred groves: some belong to tribal communities, others to the gram panchayats, some are managed by tribal elders, others by the forest or revenue department.[44] In the 1960s in Maharashtra, a Paschim Maharashtra Deosthan Prabodhan Samiti was formed to manage sacred groves.[45] Two hundred and twenty-three such groves were documented by Gadgil and Vartak.[46] In Kerala, the sacred groves may be owned by a joint family, a nuclear family, a caste group or a trust. Usually, about one-seventh of the landholding is earmarked for the maintenance of the grove. In the case of large Hindu temples, such as the Mannarasala Sri Nagaraja Temple, an ancient centre of pilgrimage for devotees of the serpent god Nagaraja, which is situated inside a sacred forest like all kaavus snake temples, the groves are managed by the Devaswom Boards under the overall control of the state government. In Kanyakumari district of southern Tamil Nadu, the groves are owned by Nair and Namboodiri families.

Sacred groves, as patches of virgin forests venerated on religious grounds, have preserved many rare and endangered plant and animal species, which have medicinal and agricultural value. They represent the ancient Indian tradition of in situ

conservation of genetic diversity, while Gadgil and Vartak[47] observe: 'In many parts of India, sacred groves represent surviving examples of climax vegetation.'[48] These miniature forests within human settlements are sustained by tradition and faith and linked with rituals and festivals of the local community.

The evolution of the belief system attained its zenith with the Bishnois who live in the deserts of Rajasthan. The orans at Peepasar and Khejarli villages are revered by them; the former is the birthplace of Guru Jambeshwar-ji, founder of the Bishnoi cult, and the latter symbolizes the supreme sacrifice of 363 Bishnois who protested the felling of the khejri tree in Khejarli village in the year 1730 CE by the subjects of the Jodhpur king. If the khejri plant, deer, blackbuck and nilgai and other animals of the Thar Desert have survived today, it is because of the commitment of the Bishnois, for whom conservation is a religious precept. Traditional societies often use symbols and cosmologies to impart the knowledge of conservation to younger generations. This in turn helps them to conserve the existing resources intelligently.

Gadgil and Vartak[49] have pointed out the ecological wisdom found in taboos. They are like ancient strictures that have been handed down through generations. They govern the thinking and actions of the tribe or clan.[50] Taboos have thus played a useful role in preventing the destruction of the groves. These restrictions govern the social life of the community or tribe and help in preserving local ecology. They have played a very important role in maintaining the balance of nature. The fact that there are no artificial boundaries or fences for the groves is a case in point; the belief system is the 'social fencing'.

In Meghalaya it has been found from both primary and secondary sources that there are as many as 514 species representing 340 genera and 131 families; the groves are also home to many medicinal plants. Endangered species are also found here. Besides trees and shrubs, various species like lianas, orchids, ferns, bryophytes and microbes also grow in the sacred forests. The plant and animal diversity can be compared to those found in biosphere reserves. This reiterates the fact that the traditional forest management systems are more efficient than modern systems. Rightly, they can be called 'Mini Biosphere Reserves'.[51]

A grove (0.45 hectares) owned by the Gawli family of Muradhpur village in Maharashtra contains old samadhis. This is the site for performing the ceremonial rites for the deceased.

In the groves of Virachilai and Kothamangalapatti in the Pudukkottai district of Tamil Nadu, there are giant trees such as *Tamarindus indica* (17.6 m in height), *Ficus benghalensis* (15.4 m), *Albizia amara* (14 m), *Azadirachta indica* (11.9 m) and *Drypetes sepiaria* (10.75 m); all are impressively tall and robust, thereby justifying the claim that sacred groves are a museum of giant trees, gene bank of economic species and a refuge for rare and relict taxa, besides serving as spiritual retreats.[52]

Sacred groves in the Kanyakumari district support numerous rare endemic orchid species and harbour many of the rare endemic plants of the Western Ghats.[53] A miniature sacred grove measuring the size of a basketball court on the Passumari hilltop, near Vedanthangal bird sanctuary in Chingleput, Tamil Nadu, has 110 flowering plants, and is a refuge of rare species like *Amorphophallus sylvaticus*, *Kedrosiis foeiidissima*, a rare cucurbit, *Strychnos ienticellata* and the insectivorous plant, *Drosera burmanii*. A huge fig tree, about 200 years old, stands majestically in the centre. Below it are clumps of 1-m tall *Amorphophallus* plants, which flower in June–July and bear attractive red berries in August–September.

The undisturbed atmosphere and the shade provided by these trees have provided an ideal ambience for their survival and proliferation.[54]

Flying foxes (*Pteropus giganteus*), also known as the greater Indian fruit bat, play a significant role in forest ecosystems. This is especially true of the semi-orange fruit bats. They pollinate flowers and disperse seeds of trees, shrubs and climbers, all of which are a part of their functions in the ecosystem. Besides, bat droppings in caves support a delicate ecosystem composed of unusual organisms. These bats are believed to be protected by the deities associated with the sacred groves, also their roosting sites. Local people do not hunt or even allow anybody to hunt them because they believe that if they disturb the bats, the deity will punish the hunters. During marriages and other festive occasions, local people never use loudspeakers or crackers near the groves where the bats are roosting as the sounds may disturb them. There is a local belief that seeing a bat before setting out to work is a good omen and nothing is done to disturb the animal. The tree in which they roost, generally the banyan, is protected with great care. Sacred groves that are protected for their bat populations can be found in the Cuddalore, Villupuram, Thanjavur, Tiruchirappalli, Pudukkottai and Ramanathapuram districts of Tamil Nadu.

The natural heritage of sacred groves is a veritable treasure unsurpassed in any other part of the world.[55] They represent a variety of vegetation ranging from typically evergreen to dry, deciduous forest types—of the Himalayan ranges and the Western Ghats, swamps of the coastal plains along the western coast, oases of thorny scrub in the Aravalli ranges and scrub woodlands of the Coromandel coastal belt, corresponding to different climatic zones.

The temple within the groves was a later development. Temple construction was encouraged by the Mauryan kings and intensified during the Gupta period. Thereafter, it spread very fast throughout the country. Brahmanical Hinduism spread vigorously in the Western Ghats from 400 CE, and readily integrated primitive local cults; in turn, the locals were also attracted by the Hindu pantheon and adopted it,[56] especially as they came with a structured pantheon of gods and a distinct hierarchy. The deity passed through several evolutionary stages: from a stone fetish suggesting a human or animal figure to an unorganized clump of bricks/stones; a metal object, even a trident; a carved stone relief or a statue on the floor; till finally it became a figure in the round, on a platform and surrounded by an ornate temple. The mysterious eruption of a termite mound, with or without association with a tree, could be associated with the goddess. Occasionally, gods like Ganesha or Hanuman were installed besides the presiding deity. An inevitable corollary of the importance given to temples and images was the gradual neglect of the vegetation; cult practices and cultural traditions gained precedence over conservation practices. The attitudes changed slowly, shifting the primacy from plants to prayers, conservation to culture and groves to gods. And that started the nemesis of the sacred groves.[57]

There is a tradition of erecting a hero stone and worshipping it as a deity in the sacred groves of south India in memory of heroes who laid down their lives defending the grove, or a man who died in a battle with a tiger or leopard or bear within the grove, or one who made some supreme sacrifice for the sake of the local forest. Usually these stones show the figure of the hero, the battle or the king in whose time the fight took place or the person who erected the stone. Sometimes, the hero stone may be carved with inscriptions, giving details of the person and the period. Either they stand alone or in

groups and are usually found outside the village limits, near a
tank or lake or within the groves. Every grove generally has an
elaborate *sthala purana* or story to justify its special status.

Sacred Gardens[58]

Nandavanam (forests of pleasure) or temple gardens were groves
associated with ancient temples during the medieval period
in south India. There are vivid descriptions of them in about
300 epigraphs belonging to the period between the third and
fifteenth centuries CE. Several inscriptions refer to the grant of
land by rulers to maintain temple gardens called *thirunandavanam*.
Many varieties of flowering plants were cultivated and flowers
from these gardens were offered to the deity for puja.

Sacred gardens are an ancient tradition in many major
cultures, including Hinduism. They are the cultivated
counterparts of the sacred groves and are a place for
meditation, spiritual awakening and celebration. Although
there is less archaeological evidence of early gardens in India,
Hindu scriptures and books like the Ramayana, *Abhijnana
Shakuntalam, Mrichchhakatika,* etc. give remarkably detailed
descriptions of elaborate gardens with flower beds, lotus
ponds, fruit trees, creepers and shady spaces. In fact, gardens
are a symbol of paradise in Hindu philosophy and art.

There are three types of sacred gardens:

- Most Hindu temples have gardens known as
  nandavanam, associated with the *leela*s of the
  deity. These gardens are usually managed and
  maintained to serve the temple. The Madurakavi
  Nandavanam attached to the Ranganathar Temple
  at Srirangam is one such.

- In Buddhism, gardens are described as a place for meditation and healing. There were beautiful gardens in Nalanda and Taxila. Lord Buddha was born under a tree in the Lumbini garden (in Nepal), which is now listed as a World Heritage Site, and gave his first sermon at the Deer Park at Sarnath. The gardens were a central part of the life in the monasteries during the early periods, recreating an atmosphere of perfection. Even today, Buddhist monasteries in India have attractive gardens attached to them.

- Baghs (*bagicha*) are ethno-silvi-horticultural gardens, traditionally planted near tanks, settlements or amidst forests, especially in north India. The gardens mainly consist of utility trees such as *Mangifera indica, Madhuca latifolia, Syzygium cuminii,* etc. Green felling is totally banned in these gardens. There is also a temple or a separate space dedicated to the deity. For example, there is an excellent bagh near a village inside the Darrah Wildlife Sanctuary in Kota.

There is a tenth century inscription belonging to Chola king Rajaraja I which refers to a temple of Kalar, the leader of Aiyanar's army. Another inscription relates to Pidari Amman in the village of Maganikkudi in Venkonkudikandam in Maranaadu and mentions a '*nandavanam* of coconut trees'. Even today, the temple of Sri Ranganathaswamy at Srirangam near Tiruchirappalli has a floral garden called Madurakavi Nandavanam which extends over an area of 10 acres. Apart from the garden, there is also an orchard where every tree is named after a Vaishnava saint or acharya.[59]

Tiruchanur is a town in the Chittoor district of Andhra Pradesh, famous for the temple of Goddess Padmavati or Alamelumanga, the consort of Lord Venkateswara of Tirumala. The temple is administered by the Tirumala Tirupati Devasthanam. A garden, with many varieties of jasmine and other fragrant plants, covering 4 hectares, is dedicated to Goddess Padmavati. The flowers from this garden are used exclusively for decorating the goddess and in pujas.

Ornamental, landscape and flower gardens occupy an area of 460 acres in Tirupati and Tirumala, managed by the Tirumala Tirupati Devasthanam (TTD). There are about 200 varieties of plants in these gardens. It is believed that the Vaishnava acharya Sri Ramanuja and his disciple Sri Ananthalvar developed these gardens in the fourteenth century. Legend has it that the Tirumala flower gardens were cultivated by Sri Vaishnavas under the name of Dasa Nambis who made flower garlands for use in the temple. Inscriptions in the temple refer to numerous gardens during the latter period of the fourteenth and fifteenth centuries. In Tirumala, many places are named after nandanavanams—Anthalvar Nandavanam, Tharigonda Venkamamba Nandavanam, Hathiramji Nandavanam and Tallapaka Nandavanam. There are also many tanks and ponds such as Alwar tank, Mangalabhavi tank and Ananthapalligunta tank which are useful not only as perennial water sources for the temple gardens but also for growing lotus flowers.

Melkote in the Pandavapura taluk of Mandya district is a sacred Vaishnava pilgrimage site in Karnataka. It is also known as Thirunarayanapuram, the name derived from the temple of Narayanaswami which is built on a hillock, surrounded by a fort. Here, a garden is dedicated to Sri Narayanaswami, covering 6 hectares, which has about twenty-seven varieties of jasmine and a number of unusual plants and about 200 species of birds.

Odisha has several sacred gardens: Ekamravan garden, Gundicha Temple garden, Hanuman Vatika, Raja Rani Temple garden and Rani Sati Temple garden, to name a few.

Known as the garden house of Jagannath, the Gundicha Temple stands in the centre of a beautiful garden with coconut, mango, neem and *bael* trees and the favourite flowering plants of the Lord, such as tulsi, rose, jasmine, etc. The temple is occupied during the famous rath yatra of Puri, when the images of Lord Jagannath, his brother Balabhadra and sister Subhadra are brought here.

The eleventh-century Raja Rani Temple at Bhubaneswar is famous for its beautifully sculpted figures and the miniature replicas of itself decorating the spire, reminiscent of the temples of Khajuraho. It is situated in the midst of a beautiful garden with flowering plants.

Rani Sati Temple is situated at Birmitrapur, about 35 km from Rourkela. There are two beautiful flower gardens in the temple complex whose flowers are used for the daily puja.

The temples of Ranakpur are situated in the Pali district of Rajasthan. Built in the fifteenth century, the temple complex of Ranakpur is the largest of Jainism. The temples are situated within an enclosure which is treated as a garden. There is a 600-year-old sacred tree in the main temple courtyard. The temple is built around this tree. Jainism sees the natural world as governed by laws based on the interconnection of gunas (attributes) of *dravyas* (substances) which comprise the cosmos. All of nature is interdependent: if one does not care for nature one does not care for oneself. Jain temples are renowned for their sensitivity to the surrounding landscape. So trees are never cut down to build a temple.

The Pareshnath Temple of Kolkata is remarkable for its exquisite beauty. This Jain temple is surrounded by a garden with a variety of flowers, fountains and a reservoir with colourful fish swimming near the surface of the glistening water. The temple radiates peace and serenity that has a very calming effect on the mind, body and spirit.

Located about 10 km from the famous Hindu pilgrimage centre of Varanasi, Sarnath, where Lord Buddha delivered his first sermon after attaining enlightenment, is one of the most sacred Buddhist sites in India. Stupas, ruins of ancient monasteries, temples and gardens are among the prominent attractions in Sarnath. The Deer Park is the most important garden in Sarnath. The recreated garden has lotus pools and a sacred bodhi or pipal tree, grown from a sapling of the original bodhi tree at Bodhgaya under which the Buddha attained enlightenment.

Vrindavan on the Yamuna, near Mathura in Uttar Pradesh, is famous as the home of Lord Krishna. Lord Krishna and Radha once wandered through the gardens of Vrindavan. Among the most famous is Seva Kunj, where Lord Krishna performed the rasaleela dance with Radha. There is a small temple dedicated to the two of them called Rang Mahal. The other famous garden is Nidhuban, where Krishna is said to have rested with Radha after a tiring day of merrymaking. There is a shrine in the premises which houses a bed, which is decorated every morning by the temple priest in memory of their love story. Nobody is allowed in this garden after dark, as Krishna is said to frequent the place with Radha after sunset.

Krishna's association with gardens has resulted in many being named 'Vrindavan'. For example, the Madanagopalaswamy Temple in Madurai, Tamil Nadu, dedicated to Lord Krishna, has a small 'Vrindavan' within the premises, where flowers are grown and offered daily to Lord Krishna.

The Rajagopalaswamy Temple in Mannargudi in Thiruvarur district, Tamil Nadu, is also known as Champaka

Aranyam, because, in the past, it abounded with champaka trees. There is a nandavanam in the southern *prakara*, from where jackfruit is plucked to prepare a delightful dish for the ceremonial offerings to the Lord.

Oppiliappan Temple, situated near Kumbakonam in Tamil Nadu, has a garden around the third precinct of the temple and a larger garden to the south, where tulsi and flowers are grown.

The temple of Varadaraja Perumal in Kanchipuram has a nandavanam behind the shrine to Sudarshana Chakra (also known as Chakrat-Alvar). It is called Dorai Thottam. The garden has a tulsi vanam, coconut, mango and champaka trees and different kinds of flowers used for offering the deity.

Sri Ranganathaswamy Temple, located at Srirangam, near Tiruchirapalli, Tamil Nadu, has a sacred garden or nandavanam dedicated to Lord Sri Ranganatha, which covers 4.05 hectares. The garden has an orchard where every tree is named after an Alvar, a Vaishnava saint of the Tamil tradition. Lord Ranganatha and his consort are decorated with flowers and only from the Madurakavi Nandavanam. Sri Ramanuja, the Vaishnava saint, gave an important position to the gardeners by creating the Dasa Nambis. They laid out the gardens, provided the flowers and garlands and decorated the palanquins with flowers for the processions. The Madurakavi Nandavanam is tended by the Ekangis in charge, who follow a strict regimen of personal purity and ensure that the garlands-in-making do not touch the ground but are kept on a bench. Every year more than 2000 garlands are strung from the flowers grown here and sent to the temple. Flowers are woven into shawls, crowns, flower chariots (*poonther*), etc. by the deft fingers of the garland makers of Srirangam.

Srivilliputtur, situated in the Virudhunagar district of Tamil Nadu, is the birthplace of the Vaishnava saint Peri

Alvar and his foster-daughter, Andal, the only female Alvar who probably lived in the eighth century. There are two temples in this town: one of Vatapatrasayi

Figure 2.3: Madurai temple garden

Source: C.P.R. Environmental Education Centre, www.cpreecenvis. nic.in

and the other of Andal, which is believed to be the original residence of Peri Alvar. It is locally known as Nachiyar Thirumaligai. On the way to the Andal shrine is the Tirupura Nandavanam, established by Peri Alvar, an important landmark of this temple, for Andal was found as a baby inside the nandavanam. It is a small shrine with an image of a smiling baby and a tulsi mandapam beside her. Andal was found as a baby here by Peri Alvar on Adi Pooram day.

Koodal Azhagar Temple in Madurai, now overgrown with weeds, is an example of the sad state of neglect of most nandavanams. The world-famous Meenakshi Amman Temple, also in Madurai, has a garden where flowers are grown to be offered to the deity (Figure 2.3).

Nature being an indivisible part of Hinduism, ancient Indians saw in spirituality a method to protect their forests and biodiversity, especially those plants and animals essential for their lives and existence. Thousands of plants and animals of the groves were held sacred in the tapovanas, especially as rulers and their insatiable appetites for timber and other forest produce, as well as their predilection for hunting, made unmanageable demands on the forest, not unlike today. The taboos against harming the groves would have frightened the kings, who were both religious and superstitious. To ensure the survival of important species and their forests, local people invoked the gods, especially the Earth Goddess or Bhoomi Devi of the *Atharva Veda*, who is the reigning deity in most groves all over India.

Sacred groves also provide a window into the concerns of rural and tribal communities, the plant and animal species they protected and the elaborate mythologies they created to ensure this. Their existence is very important for a study of India's environmental and forest history, for they indicate the species of plants and animals that local people believed were important, and the methods used to preserve them.[60]

The deities of the groves reflect the ecological concerns of local people. The conservation was carried out in spite of the many threats of deforestation in the colonial period and in post-Independence India. In many places, the sacred groves are the last remnants of local biodiversity. Because of the myths created by tribal and rural communities all over India to protect their forests, we are still able to have a glimpse of what the great forests—the mahavana—of ancient India looked like and their biological diversity, unlike nations of Europe who have lost their sacred groves. In spite of the reforestation in Europe in the last one hundred years, they have only been able to create mere plantations: the native species, which are an essential part of the forest, are gone forever.

Today, the fundamental concept of sacred groves—a traditional belief system—is threatened by developmental activities, agriculture and urbanization. The violation of religious norms and taboos no longer carries any consequences. The destruction of sacred groves is a violation of the rights of the indigenous and rural people, whose lives depend on them.

The Bishnois of Jodhpur district, Rajasthan, as we have seen, are the best examples of community protection for sacred groves which has led to the conservation of several thousands of orans.

3

Divine Waters

Water is the elixir of immortality.
From the waters is this universe produced.
In the waters, O Lord, is your seat, that is, in the waters
  O Lord, is your womb . . .
The waters are the foundation of all this universe.

—*Shatapatha Brahmana,* VI.8.2.3–13

W hen the ocean was churned for the nectar of immortality, a number of items came forth from the waters. Among them was Lakshmi, the goddess of prosperity and apsaras, the divine nymphs who were a class of demigoddesses. Thus the waters were, in more than one way, a source of prosperity and good fortune.

Water is sacred because all life depends on it: it is the source of survival and energy, the medium of self-purification. 'The waters in the sky (rain), the water of rivers, and water in the well . . . may all these sacred waters protect me,' sang the poet of the *Rig Veda* (VII.49.2). The mountains were sanctified by their association as the source of rivers and lakes, and the homes of the gods themselves. Many pilgrimage sites are found on riverbanks; sites where two or even three rivers converge are considered particularly sacred.

There are many sacred rivers: the Ganga, Godavari, Kaveri, Narmada, Sarasvati, Sindhu, Yamuna and Brahmaputra. They are all supreme but the Ganga is the most sacred of them all, assuring the bather forgiveness for all sins committed in this and previous births. According to Hindu belief, those who bathe in the Ganga or who leave part of themselves (hair, bones of the dead) on the left bank of the river will reach *swarga*, the paradise of Indra, the god of the heavens. In fact, bathing in any river is a sacred act and assures the bather instant salvation.

The *Rig Veda* praises other rivers in the 'Nadistuti Sukta' (hymn in praise of rivers) (X.75). The ten rivers are listed, beginning with the Ganga and moving westwards: 'O Ganga, Yamuna, Sarasvati, Shutudri [Sutlej], Parushni [Iravati, Ravi], follow my praise! O Ashkini [Chenab], Marudvridha, Vitasta [Jhelum], with the Arjikiya [Haro] and Sushoma [Sohan], listen'.

The next stanza includes north-western rivers flowing through modern Afghanistan and north-western Pakistan: 'First united with the Trishtama in order to flow, with the Susartu and Rasa, and with this Svetya [you flow], O Sindhu [Indus] with the Kubha [River Kabul] to the Gomati [Gomal], with the Mehatnu to the Krumu [Kurram], with whom you rush together on the same chariot.'

Funeral rites are always held near rivers; the son of the deceased pours water on the burning funeral pyre so that the soul cannot return. When the fire reaches the deceased's skull, the mourners bathe and then go home. The ashes are collected three days after cremation and, several days later, immersed in a holy river.

The myth of the Great Flood is repeated in several Hindu texts, which tell the tale of how Manu, the first man, was

rescued from the flood by a fish (Brahma), who led him to the
Himalayas until the waters receded.

The earliest civilizations are found along the banks of rivers.
Ganga is worshipped as a goddess, with her own iconography,
legends and temples. She comes cascading down the hills from
Lord Shiva's topknot. Even her presence on earth is a divine
response to the penance of King Bhagiratha.

The rishis of the *Rig Veda* called the waters goddesses, for
they quenched the thirst of their cattle (I.23.18). They explain
why the rivers should be praised:

> Amrita is in the waters; in the
> Waters there is healing balm: be swift,
> Ye gods, to give them praise,
> Within the waters – the waters hold all medicines.
> (I.23.19–20).

Apsaras are the anthropomorphic form of the waters. They are
mothers or young wives; they flow in channels to the sea; but
they are also celestials.[1]

In literature, all the rivers of India attained sacred status.
This stanza says it all:

> *Gange cha Yamune chaiva Godavari Sarasvati*
> *Narmada Sindhu Kaveri jale asmin sannidhim kuru.*

which means,

> O sacred Ganga, Yamuna, Godavari and Sarasvati,
> Narmada, Sindhu and Kaveri, please be present in this
>   water beside me (and make it sacred).

Water is imbued with powers of spiritual purification for
Hindus, for whom morning cleansing with water is a daily

obligation. All temples are located near a water source, and followers must bathe before entering the temple. The body and mind get equally purified during a bath. Ideally, one should undergo this purificatory ritual in a sacred river, especially the Ganga. One can also bathe in any major river, all of which are sacred. And finally, if one is not beside a river, the above sloka can invoke all the rivers in any bucket of water.

Sarasvati

Sarasvati was the most important of the Vedic rivers: 'Oh Mother Sarasvati you are the greatest of mothers, greatest of rivers, greatest of goddesses. Even though we are not worthy, please grant us distinction' (*Rig Veda*, II.41.16). There are several Rigvedic verses invoking her as a mother, a great purifier. She is described as flowing into the *samudra* (sea or ocean): 'This stream Sarasvati with fostering current comes forth, our sure defence, our fort of iron. As on a chariot, the flood flows on, surpassing in majesty and might all other waters. Pure in her course from mountains to the ocean, alone of streams Sarasvati hath listened. Thinking of wealth and the great world of creatures, she poured for Nahusha her milk and fat' (VII.95.1–2).

Forty-five of the Rigvedic hymns shower praise on the Sarasvati. No other river or geographical feature comes close in importance. The Ganga is barely mentioned twice and the Indus, although referred to as a 'mighty river', is not given the same reverence. In contrast, the Sarasvati is called the mother of all rivers and 'great among the great, the most impetuous of rivers'. She is even called the 'inspirer of hymns', suggesting that the *Rig Veda* was composed on its banks. Unlike later texts, it does not mention a dying Sarasvati. Instead, it mentions clearly that she entered the sea in full flow. This would suggest that the text was composed before 2600 BCE,[2] when the river is believed to have gone underground or ceased to flow.

The first reference to the disappearance of the river is to be found in the *Jaiminiya Brahmana* (II.297), which speaks of the diving under (*upamajjana*) of the river. According to the Mahabharata (III.88.2), the Sarasvati dried up in the desert, and the place where she disappeared was visited by Balarama.

In the *Rig Veda* (X.30.12), Sarasvati is described as a river goddess and invoked as a protective deity, which may explain her appearance in later Hinduism as a goddess of knowledge, learning, wisdom, music and the arts. The river goddess became the goddess of knowledge in the later Brahmanas which identified her as Vagdevi, the goddess of speech, perhaps in memory of the Vedic religion which evolved on the banks of the river, and where oral study of the Vedas was essential to acquire knowledge of literature, philosophy and the arts. The Rigvedic River Sarasvati is today identified with the Ghaggar-Hakra. Even after the river disappeared, either due to seismic activity in the region or climatic changes,[3] she is still celebrated as the goddess of learning and the arts in India. There are many communities of Saraswats found all over India, named after Sarasvati, who trace their origins to this lost river.

The Indus and Its Five Tributaries

The Indus—Sindhu in Sanskrit—is a major river and one that has given Hinduism its name. The Sapta Sindhu were known to the Persians as Hapta Hindu and the people who lived in the region of the Hind (later Hindustan) became Hindus. The Greeks called it River Indos, from which came the word India. Originating near Mount Kailas and Lake Manasarover in Tibet, the river enters India in Ladakh. Many sites of the Indus–Sarasvati civilization—especially Mohenjo-Daro and Harappa—were situated along its banks. Greek accounts of the Indus Valley at the time of Alexander's campaign indicate that there was a rich forest cover, which was reduced due to extensive deforestation. There was a thick forest here when Mughal king Babur came to India, for he went hunting rhinos along the Indus, according to his memoirs (*Baburnama*). The massive deforestation led to the deterioration in soil quality and vegetation, making agriculture dependant on irrigation in a land that was once a rich forest. Unfortunately, the waters of this sacred river came under the territory of Pakistan after Partition.

In the Rigvedic hymns, all the rivers are feminine in gender except the Sindhu. While other rivers are goddesses and are compared to cows and mares yielding milk and butter, the Sindhu is described as a strong warrior.

Punjab is in the region of the Sapta Sindhu mentioned in the *Rig Veda*. The present tributaries of the Indus are the Beas (Vipasa), Sutlej (Shutudri), Jhelum (Vitasta), Chenab (Ashkini) and Ravi (Iravati).

The Sanskrit Vipasa or modern Beas is a famous river of Punjab. Rishi Vashishtha was stricken with grief at the death of his beloved son, Shakti, and jumped into the river with a rope in order to commit suicide. But the waters of the river untied the knots of the rope and saved him. River Beas marks the eastern border of Alexander the Great's conquests in 326 BCE.[4] The name 'Beas' is believed to be a corruption of the word Vyas, the name of Veda Vyasa, author of the Mahabharata. The river begins at the Rohtang Pass in Himachal Pradesh and merges with the Sutlej at Hari-ke-Pattan, south of Amritsar. Thereafter, the Sutlej continues into Pakistani Punjab and joins the River Chenab to form Panjnad, which joins Indus at Mithankot. The waters of the Beas is allocated to India under the Indus Waters Treaty between India and Pakistan.

The Sanskrit Shutudri, better known as the Sutlej, is the longest tributary of the Indus, commencing at Lake Rakshastal in Tibet and cutting through tough Himalayan terrain before entering the plains. Sutlej played a major role in the ancient civilizations of Tibet. Its valley was called Garuda Valley, where you can still see the ruins of the famous Kyunglung Palace. Tibetans believe that the river actually originates in Lake Manasarovar. It was once a tributary of River Sarasvati.

The Jhelum's Sanskrit name is Vitasta. Goddess Parvati was requested by sage Kashyapa to purify Kashmir from the impurities of the *pishachas*. She assumed the form of a river and was released by Lord Shiva who made a stroke with his trident. Shiva named her Vitasta, a measure of the length through which the river had come out. If anybody bathes in the river and fasts for seven days it is believed that he will become as pure as a hermit.[5]

Alexander the Great and his army crossed the Jhelum in 326 BCE and met King Porus at the battle of Hydaspes—the name Greeks gave to the Vitasta. According to Arrian (*Anabasis*, 29), Alexander built a city 'on the spot whence he started to cross the river Hydaspes', which he named Bucephala, after his famous horse Bucephalus. It is the westernmost of the five rivers.

River Chenab was Ashkini or Chandrabhaga in Sanskrit. Alexander the Great founded the town of Alexandria-on-the-Indus on the Chenab or Greek Acesines at the confluence of the Indus and the combined stream of the Punjab rivers. It originates in the Lahaul and Spiti district of Himachal Pradesh and flows through Jammu into the plains of Pakistani Punjab.

The Battle of the Dasharajna (ten kings) is a famous war in the *Rig Veda*. The Ravi, which was known as Parushani or Iravati to Indians in the Vedic period, was the site of this battle. It was fought on 'a river' which, according to Yaska (*Nirukta*, 9.26), is the River Iravati. Divodasa, the fifth successor of Sudasa, defeated his foes on the Yamuna and again defeated Puru and others in battles on the River Parushani or Iravati (modern Ravi). Hence he must have driven Puru out of the Paurava kingdom of Hastinapura first to the Yamuna and then as far west as the Ravi.[6]

Ganga

The Ganga is the most sacred river of India. Millions of Indians who live along its banks depend on it for their survival and livelihood. Many former imperial capitals—Pataliputra (Patna), Kashi (Benares), Kanauj and Kolkata—are situated on its banks. As the cities of the Indus–Sarasvati civilization declined, and River Sarasvati gradually disappeared or ceased to flow, the civilization of the Vedic people moved eastwards. Indian civilization shifted from the Indus to the Ganga around 2000 BCE.

River Ganga flows from the Himalayas to the Bay of Bengal. She is known as a goddess and has a prominent position in the culture of Hinduism. The Ganga is central to the Hindu way of life. For several millennia, people have settled on her banks; her rich soil is most conducive to agriculture, making her the rice bowl of north and east India. Her water is sacred and cleansing, and is used for drinking and irrigation. She cleanses the sins of human beings. Her existence is divine.

The Ganga is revered all along her length. The most sacred pilgrimage centres are found along her course:

Gangotri, Uttarkashi, Rishikesh, Haridwar, Kashi, Kanauj and Allahabad, where it combines with the Yamuna and the Sarasvati, which is believed to flow underground, to meet at Triveni Sangam in Allahabad. By praying in her waters, one is purified of one's sins. The Ganga takes one's prayers to the ancestors; the devotee prays by facing the sun and holding the water in his cupped hands and then letting it fall back. The Ganga *aarti* every evening and the shallow clay dishes, filled with oil and lit with wicks, that float by are a part of the mystical beauty of the Ganga. Flowers are offered to the river as a prayer offering. Visitors carry back its waters in small sealed brass pots or even plastic cans. It may be used for rituals, or as a deity to be worshipped, or even stored and used to wash the dead body of a beloved relative. Finally, in death, the ashes of the dead person are immersed in the river. While several streams join the Ganga, the six longest—Alakananda, Dhauliganga, Nandakini, Pindar, Mandakini and Bhagirathi—are considered sacred.

Every twelve and six years alternately, the Kumbh Mela is celebrated at Prayag in Allahabad. The first mention of it is by the Chinese scholar Hsuan Tsang who visited India between 602 and 664 CE.[7] According to legend, the gods once lost their immortality and had to churn the ocean of milk for nectar or *amrita*. They were forced to ask their foes, the asuras or demons, for help. This is the famous story of the *samudra manthana* which is associated with Vishnu's Kurma avatar. The demons did not notice Vishnu's mount Garuda fly down and carry off the *kumbh* (bowl) containing the nectar or amrita. It fell at four places: Allahabad, Haridwar, Nashik and Ujjain, which are the sites of the mela. On this day, it is believed, the amrita is present in the waters of the river. Every Hindu hopes to bathe in the river at Allahabad during the Kumbh Mela.

Varanasi or Kashi (Figure 3.1) is the most sacred city of the Hindus and Jains and is closely associated with the history of Hinduism, Buddhism and Jainism. The ghats of Varanasi are the sites of a magnificent spectacle that breaks at sunrise and ends late at night. Worshippers enter the sacred waters at dawn and the cremation of dead bodies goes on through the day. After the sun sets, the famous Ganga aarti begins, while the devout set lit diyas or clay oil lamps to float down the river—till the city goes to sleep. There is a force and energy about Kashi which entices and powers every visitor to this day.

Figure 3.1: Ganga

The myths surrounding the Ganga are endless. She is believed to have been the spouse of King Shantanu, the ancestor of the Kurus, who happened to see her in her human form. Overwhelmed by her beauty, he asked her to marry him. She agreed, but made the condition that he should never question her actions. If he did, she would leave him. Shantanu agreed, and the two were married. When their first son was born, Ganga took him to the river and drowned him. Shantanu was horrified, but his promise forced him to keep quiet. This happened to six more children. When the eighth, a son, was born, he could no longer contain himself and demanded to know the reason. Ganga then informed him that they were cursed Vasus (eight minor elemental deities representing the aspects of nature; Vasu means 'dweller' or 'dwelling', whose souls she had liberated). She then took her eighth son, Devavrata, with the promise that she would return him on his sixteenth year. This child was later known as Bhishma, an important character in the Mahabharata.

The *avatarana* or descent of the Ganga from the heaven to the earth is celebrated every year in the month of Jyeshtha. Ganga Dashahara takes place on the tenth day (*dashami*) of the waxing moon. Bathers throng the banks of the river. A dip in the river on this day is said to rid the bather of ten sins or ten lifetimes of sins. Those who cannot go to the river can achieve the same result by bathing in any nearby river or waterbody, which takes on the sanctity of the Ganga.

It is Shiva, however, among the major deities of the Hindu pantheon, who appears in the most widely known version of the avatarana story.[8] The description of the descent of the Ganga appears in the Ramayana, Mahabharata and many Puranas. The story goes that sage Kapila's meditation was disturbed by the sixty thousand sons of King Sagara. Furious, Kapila reduced them to ashes. Only the waters of the Ganga could bring them back to life. King Bhagiratha, a descendant of Sagara, undertook a rigorous penance, and was granted his wish by Lord Shiva. Ganga would descend to the earth. However, she was so angry that she left the comfort of heaven and came down with a great fury, threatening to destroy the earth. Bhagiratha again prayed to Shiva to collect Ganga in his locks and allow only a small part to reach the earth. So she reached the earth in the Himalayas, from where Bhagiratha led her to Rishikesh in the plains. In honour of his role in bringing Ganga to the earth and reviving his ancestors' lives, the source river of the Ganga is named Bhagirathi. There is also a belief that she first descends upon Mount Meru, before she flows on earth.

Ganga, who descended to the earth, is also the vehicle of ascent, from earth to heaven. As the *triloka-patha-gamini*, (*triloka*: three worlds, *patha*: road, *gamini*: one who travels), she flows in heaven, earth and hell; she is, therefore, a teertha, or crossing, of all beings, the living and the dead. The story of the avatarana is narrated at *shraddha* ceremonies for the deceased. Death in Varanasi assures the person of a full cleansing of all sins and a place in heaven. Those who are lucky to die here are cremated on the banks of the Ganga and granted instant salvation. If the death has occurred elsewhere, salvation may be achieved by immersing the ashes in the Ganga.[9] The waters of the Ganga are believed to purify sins of even the most unrepentant sinner. Moving water is considered purifying in Hindu culture because it both absorbs impurities and takes them away. And the Ganga wipes away the sins of a lifetime.[10] She descends to the earth to make the land fertile and wash out the sins of human beings.

Shiva is depicted in art as Gangadhara, the bearer of the Ganga, with Ganga coming out of a spout of water, issuing from his hair. Ravi Varma's

Source: The C.P. Ramaswami Aiyar Foundation

Figure 3.2: *Descent of Ganga* by Raja Ravi Varma

famous painting *Descent of Ganga*
(Figure 3.2) illustrates this event so
beautifully.

The Ganga appears in early art
standing on a crocodile, her vahana or
vehicle. The *makara* is symbolic of the
Gangetic crocodile which once lived
in her waters. The first appearance
of the two together is in the Varaha
cave of the Gupta period at Udayagiri
(near Bhopal). In her hand, Ganga
holds the *poorna kumbh* or a full pot
of water; again, this first appears in
the same sculpture at Udayagiri. She
stands beneath the branch of a tree at
Udayagiri, which is later replaced by a
chhatra or umbrella (Figure 3.3).

Figure 3.3: Goddess
Ganga standing on a
crocodile, Gupta, 5th
century CE, National
Museum, Delhi.

The Mekong, which flows through
China, Myanmar, Laos, Thailand, Cambodia and Vietnam, was
regarded as sacred in the ancient Hindu kingdoms of South-East
Asia, so they named it Ma Ganga, from which comes 'Mekong'.
The sacredness of the Ganga has spread far and wide.

Yamuna

If Ganga's vehicle is the crocodile, River Yamuna's is the
tortoise. The name comes from Yami, the sister of Yama.
Yami means twin, an appellation that probably comes from
the fact that the river runs parallel to the Ganga, which it joins
at Triveni Sangam in Allahabad.

Yamuna was born of Surya the sun, and his spouse, Samjna.
Unable to tolerate his brilliance and heat, Samjna would keep

her eyes shut in his presence. Surya would scold her and then she would try hard to keep her eyes open, but her best efforts resulted in a mere flutter. Finally, Surya proclaimed that his son would be Yama, meaning restraint, and daughter, Yami or Yamuna. The pair of twins, being the children of Surya, were divine themselves. The poison of Kaliya the serpent defeated by Krishna made the river dark, hence her other name Kalindi.

Most of the events in the Mahabharata took place on the banks of and around the Yamuna, which is also closely involved with the story of Lord Krishna. Krishna was born in Mathura, situated on the banks of the river. On a stormy night, his father, Vasudeva, crossed the river to take him to the house of Nanda and Yashoda in Vrindavan. Krishna grew up on the banks of the river in Vrindavan where he played with his friends and danced with the cowherdesses. He fought several demons, especially Kaliya, along the banks of and in the River Yamuna. Later, as an adult, he helped his friend Arjuna build Indraprastha—in a very fiery episode—on the Yamuna; today this is where the city of Delhi is located.

Mughal history also played out on the banks of the Yamuna. The Taj Mahal was built in Agra, situated on the banks of the river.

Yamuna stands on a tortoise. Her first appearance in art is in the Gupta Varaha cave at Udayagiri (near Bhopal), holding the poorna kumbh or a full pot of water. She stands beneath the branch of a tree, which is later replaced by a chhatra or umbrella (Figure 3.4).

Figure 3.4: Goddess Yamuna standing on a tortoise, Gupta, 5th century CE, National Museum, Delhi.

The river, 1376 km long, originates in the Saptarishi Kund, a lake above Yamunotri in Uttarakhand, and passes through Himachal Pradesh, Delhi and Uttar Pradesh till it reaches Allahabad. In the sixteenth century, the saint and religious leader Vallabhacharya, a great devotee of Lord Krishna, composed the *Yamunashtakam*. The river, he wrote, descends on earth to meet her beloved Krishna and purify the world, for Yamuna is the source of all spirituality. She alone can grant freedom, even from death. The text describes her water being the colour of Lord Krishna himself, who is *shyama* (dark).[11] This is probably a throwback to the Rigvedic description of the river as *krishna* or dark. Megasthenes, in his *Indica,* refers to the river as the land of Sourasenoi (Surasena), whose capital was at Methora (Mathura),[12] which corresponds to modern Vraj.

Yamuna is, today, one of the most polluted rivers in the world, especially around New Delhi, due to the high population growth and rapid industrialization. Delhi dumps about 58 per cent of its waste into the river and there is 100 per cent urban metabolism of River Yamuna as it passes through the National Capital Territory (NCT).[13]

'A division bench of the Uttarakhand High Court has declared, in a landmark judgment, that the rivers Ganga and Yamuna, all their tributaries, streams, and every natural water flowing with flow continuously or intermittently of these rivers, as juristic/legal persons/living entities having the status of a legal person with all corresponding rights, duties and liabilities of a living person. This is the first time in India and second time in the world that rivers have been recognized as a living entity with its own rights and values and given the status of a legal/juristic person.'[14]

Godavari

The source of the Godavari is the Western and Eastern Ghats, although it is believed to begin from Tryambakeshwar in Maharashtra. It is the second longest river in India after the Ganga. It covers the states of Maharashtra, Telangana, Andhra Pradesh, Chhattisgarh, Madhya Pradesh, Odisha, Karnataka and Yanam (Puducherry), a distance of 1465 km. It is the longest river in peninsular India and is known as the Dakshina Ganga or the River Ganga of the south. Nashik, where the nectar of immortality fell, is a site of the Kumbh Mela situated along the river.

The Godavari is closely associated with the exile of Rama. Rama, Lakshmana and Sita lived on the banks of the river at Panchavati and it is from here that Ravana abducted Sita.

Sage Gautama, whose wife, Ahalya, was released from her curse by Rama, lived in the Brahmagiri Hills. He had received the boon of a bottomless grain-supplying well, but his enemies led a cow into his granary. As Gautama chased away the cow, she fell down and died. To expiate his sin, he prayed to River Ganga to descend, visit and cleanse his ashrama. Finally, the goddess yielded: she came down as Godavari along with Shiva as Tryambaka. The Tryambakeshwara Temple is one of the twelve *jyotirlinga*s of Lord Shiva.

Paithan on the Godavari, famous for its silk saris, is the location of the Digambar Jain Atishay Kshetra.

Nanded on the Godavari is associated with the first as well as the last of the Sikh gurus. Guru Nanak passed through the town, while the tenth Guru Gobind Singh came here to proclaim himself the last living guru of the Sikhs.

The river also flows through Bhadrachalam, the home of Sant Bhadrachala Ramdas.

It is equally associated with Shiva as Veerabhadra. Shiva was so furious when his wife, Sati, burnt herself to ashes in her father Daksha's yajna that he danced the *rudra tandava* in anger, creating Veerabhadra out of a lock of hair, which fell during the dance. Veerabhadra destroyed Daksha's ritual, killed him and threw his sword into the Godavari. This place is celebrated as the Veerabhadraswamy Temple in Andhra Pradesh, where a *swayambhu* (self-created) Shiva linga appeared.

Every twelve years a massive *pushkaram* or the worship of the river takes place at Rajahmundry in Andhra Pradesh. It is a combination of ancestor worship, religious discourses, bhajans and other cultural activities.[15]

Brahmaputra

The Brahmaputra is the major male river of India. It means 'son of Brahma' in Sanskrit. The river begins its journey in Tibet, where it is known as Tsangpo (purifier), originating from a height of 5150 m. It takes different names as it flows through Tibet: Tamchok Khambos Chorton (the river that gushes from the mouth of the horse) south of the Kailas range from the Chemayung Dung Glacier; Mutsung Tsangpo; Siang; Dihang; and Moghung Tsangpo. The Chinese refer to the river as Yarlung-Tsangpo-Brahmaputra (sky river). It meets the Lohit and Dibang in Arunachal Pradesh and flows into Assam as Brahmaputra. But it is also called the Lohit. The Mahabharata and Kalidasa in his *Raghuvamsha* refer to the river as Lauhitya. The name Brahmaputra occurs for the first time in the *Kalika Purana* and the *Yoginitantra*.[16]

According to the *Kalika Purana*, a sage named Shantanu lived with his wife, Amogha, on the banks of Lake Lohita. Impressed by their piety, Lord Brahma decided that Amogha was the chosen person to give birth to his son who would benefit humanity. Brahmaputra, according to the *Kalika Purana*, is the son of this couple. Amogha bore Brahma's son, who assumed the form of a large mass of water where gods and apsaras could bathe. Shantanu placed him in the middle of four mountains: Kailas, Gandhamadana, Jarudhi and Sambaka. This story explains why the Brahmaputra is considered to be a male river.[17] The lower reaches of the Brahmaputra are sacred to the Hindus.

The Brahmaputra floodplains are dotted with wetlands or *beels*, which provide unique habitats for a variety of flora and fauna. They also retain floodwaters. The Brahmaputra provides a wonderful habitat for a variety of flora and fauna, including many endangered species such as the one-horned rhinoceros, pygmy hog, hispid hare, Asiatic elephant, clouded leopard, marble cat, golden cat, binturong, hoolock gibbon and the wood duck.[18]

Later, Parashurama dug out a channel from this lake to benefit people and thus the Brahmaputra began to flow. There is speculation as to whether this lake was the original Manasarovar or the Parashurama Kund in Arunachal Pradesh.

The river originates in the Angsi glacier, north of the Himalayas in Tibet. It flows south-west through Assam as the Brahmaputra and south through Bangladesh as the Jamuna (not to be confused with the Yamuna). It merges with the Ganga, known as the Padma, in Bangladesh, and finally joins the Meghna till it reaches the sea. The Assam government celebrates the might and beauty of the river with the Namami Brahmaputra festival.

The other male rivers are the Damodar and Ajay, which flow across West Bengal and Jharkhand; the Sankosh, which originates in Bhutan and meets the Brahmaputra in Assam; the Bhima, which flows through Maharashtra, Karnataka and Andhra Pradesh before it meets the Krishna; and the Krishna, whose delta is one of the most fertile.

Narmada

The Narmada begins its journey at Amarkantak, a beautiful forested area in Madhya Pradesh and flows for 1300 km. It is one of the sacred rivers of India, invoked in the purification ritual mentioned before. It is the location of some exquisite scenery—the Kapildhara Falls, the Dhuandhara Fall of mist; the deep, narrow channel flanked by the Marble Rocks near Jabalpur; and the many waterfalls along its route—till it reaches the Arabian Sea in Gujarat. It has been the subject of many legal battles for and against the Sardar Sarovar Dam in Gujarat. The Narmada Main Canal is the largest lined canal in the world. Another major dam is the Indirasagar Dam, also in the state of Gujarat. India's major forests of teak and sal grow along this river. The Kanha and Satpura National Parks are rich in wildlife, including the elusive tiger.

The river is mentioned by Ptolemy in 200 CE and by the author of *Periplus of the Erythraean Sea*.

There are several origin stories but the most accepted one is that associated with Lord Shiva. Once there was a terrible drought on earth. Shiva meditated for several days to end the drought. So intense was his meditation that he began to sweat profusely. From the droplets of his sweat appeared a beautiful young girl whom he named Narmada (or one who gladdens the hearts of men). She was asked to provide salvation to the

human race. At Amarkantak, the
goddess has been immortalized in
the Narmadeshwar Temple. Seated
on her crocodile mount, the four-
armed goddess holds a Shiva linga,
signifying her devotion to her
creator (Figure 3.5). It is believed
that Ganga visits Narmada once a
year as a black cow to cleanse herself
of all her sins. Devotees perform a
difficult Narmada parikrama, which
is a 2600-km journey from Bharuch
to Amarkantak and back to Bharuch

Figure 3.5: Goddess Narmada

again.[19] Omkareshwar on the Narmada is the site of a famous
Shiva temple, which is one of the twelve jyotirlingas.

The river meets the sea in the state of Gujarat. Ankleshwar
and Bharuch are on opposite banks of the Narmada. Bharuch
is the centre of one of India's major festivals, the Chhath Puja,
which is a prayer of thanks to Surya, the sun god, and his
spouse, Usha, for life on earth and for granting the prayers of
the devotees. It involves four days of fasting and praying to the
rising and setting sun.

Kaveri

The Kaveri is also known as Dakshina Ganga. She has been the
lifeline of the south, although that position is slowly eroding as
dams and interstate battles vie for her waters.

Her origin is described in the *Skanda Purana* (11–14).
During the samudra manthana, Vishnu took the form of the
beautiful Mohini, in order to distract the asuras and enable
the devas to take the divine nectar or amrita. Lakshmi sent

Lopamudra, an apsara, to assist Mohini. After the amrita was restored by the devas, Lopamudra was raised by Brahma as his daughter. After some time, King Kaver, who became a sage, came to Brahmagiri Hills to meditate. As sage Kavera, he prayed to Lord Brahma to bless him with a child. Pleased by his devotion, Brahma gave him the apsara Lopamudra, who was renamed Kaveri, as a daughter. She prayed to Lord Brahma, on Brahmagiri Hills, that she may flow as a river to rid people of their sins and bless the land with fertility. Brahma granted her both wishes. But sage Agastya saw Kaveri and asked her to marry him. She could not refuse the sage, but made him promise that if he ever left her alone for too long, she would leave him and go her way. Agastya promised, but one day he was so busy with his disciples that he forgot the time and his promise. Kaveri waited for a while but when he didn't come, she jumped into Agastya's *kamandalam* and flowed out of it, as a river. Agastya's disciples tried to stop her, but Kaveri went underground and reappeared at Bhagamandala, after which she flowed southwards till she finally reached the Bay of Bengal.[20] She has been worshipped as a sacred river since that day.

There is a belief that River Ganga joins Kaveri underground once a year, during the month of Tula, in order to cleanse herself of the pollution caused by the sinners who bathe in her waters. Kaveri is considered to be as sacred as the Ganga, with the power to wash off one's sins.

The Tala Kaveri festival is celebrated at Tala Kaveri, in the dense forests of Kodagu—anglicized as Coorg—where she is believed to have originated. She is worshipped here along with Iguthappa, an incarnation of Lord Subrahmanya. After Kaveri goes underground at Tala Kaveri, she reappears to join her sister, Kannige, at Bhagamandala, along with the mythical Sujyoti, who comes from beneath the ground. The Triveni

Sangam, where the three rivers meet, is Bhagamandala, a sacred teertha on River Kaveri. There are several tributaries that join the Kaveri on her journey. Several rivers and rivulets have their origin in the Sahyadri range and many of them— Bhavani, Kapila, Lakshmana Teertha, Hemavati, Ekadi, Harringi and Kabini—join the Kaveri. Copious rainfall in the Sahyadri is necessary for the delta region on the east coast to remain fertile. In Kodagu, the river cuts her way through hills, passing luxuriant vegetation all along the way. In the Chola kingdom, Kaveri was known as Ponni (the golden river). Today people associate the name Ponni with a popular variety of rice named after the river and part of her golden bounty! In poetry, she is eulogized as the one who brings golden soil and bestows golden harvest. Early Tamil literature describes her banks as being covered with areca nut, coconut, jackfruit, mango, plantains, sugar cane and paddy. Poets also talk of ivory, coral and pearls brought down by the river.[21]

There are so many temples along the river, making it one of the most sacred lands. The Pancharanga Kshetrams (five *arangam*s or Ranganathas) is a group of five sacred temples on the Kaveri dedicated to Ranganatha, a *shayana* (a resting/ sleeping) form of Narayana. The five are Srirangapatna, called the Adi Ranga, the first temple on the banks after Tala Kaveri; Srirangam, the island formed by the Kaveri near Tiruchirappalli; Appalarangam at Tiruppernagar; Parimala Ranganatha at Tiruindalur in the Mayiladuthurai district; and the Vatarangam near Sirkazhi.[22] The first is in Karnataka and the last four are in Tamil Nadu. But the temples begin at Tala Kaveri and Bhagamandala, going on to Shivasamudram, all in Karnataka. The ancient Someshwara Temple, visited by Adi Shankara and the temple of Ranganathaswamy or Mohana Ranga are situated at Shivasamudram.

The river enters Tamil Nadu in the Dharmapuri district. The three minor tributaries—Palar, Chennar and Thoppar—enter the Kaveri before Mettur Dam, which was built at the confluence of the Bhavani, Kaveri and the mythical Akash Ganga. According to legend, Kubera saw a tree near which a deer, a tiger, a cow, an elephant, a snake and a rat were amicably standing and drinking water from the river. It was a place inhabited by holy men and gandharvas. A voice told Kubera that the Vedas came to the earth at this place near the tree and that there was a Shiva linga beneath it. It advised him to worship Lord Shiva, who, thereafter, appeared before him. Here Shiva is known as Sangameshwarar or Amrithalingeshwarar, and the reigning goddess is Bhavani. People perform the last rites for their ancestors where the three rivers meet. When dead bodies are burnt in the Bhavani, the skulls do not scatter. It is believed that this is because there are 1008 Shiva lingas under the ground. Devotees offer rice mixed with pepper and jeera (cumin seeds) to Lord Shiva for curing illnesses and for facilitating marriages.

The Cholas built several hundred temples along the Kaveri, at Chidambaram, Nagapattinam, Mayiladuthurai, Kumbakonam, Thanjavur, Tiruchirappalli and Karur, on both sides of the river. They are also known as the 276 Paadal Petra Sthalangal, about which the Shaivite saints of Tamil Nadu sang with absolute devotion. Kumbakonam is the temple town of Tamil Nadu, with temples dedicated to Shiva and Vishnu and the Navagrahas (nine planets) dotting the entire district. The great Brihadishwara Temple of Thanjavur is situated on its banks as are many others.

King Karikala Chola constructed a bank for the Kaveri 1800 years ago, from the ancient town of Puhar (Kaveripoompattinam) to Srirangam. The *kall-anai* (stone dam) constructed by him between Tiruchirappalli and Thanjavur was made with earth and stone and survived for hundreds of years till it was renovated in the nineteenth century and renamed the Grand Anicut. This dam, one of the oldest man-made dams in the world, enriched the irrigation system of Thanjavur district. The Kaveri, which is utilized every inch of its route, is a mere trickle as she enters the sea at Poompuhar, once a fabled port town on the east coast.

Every river in India—and they are innumerable—is regarded as sacred, with a local history or legend to substantiate the sanctity.

The Sarayu is sanctified by the fact that on its banks is the town of Ayodhya where Rama was born and ushered Rama Rajya.

The capital of Vijayanagara—Hampi—was built on one bank of the Tungabhadra. The opposite bank is Kishkinda, where Rama met Hanuman and supported Sugriva in his battle against Vali. His visit sanctified the river.

Adi Shankara established the Sharada Peetham on the banks of the Tunga. Alampur on the north bank is known as Dakshina Kashi. And Guru Raghavendra created his Moola Vrindavana on the banks of the river at Mantralayam.

The Himalayas are the source of several rivers, each with its own history and mythology. The melting snows give rise to the rivers Alakananda, Gandaki, Teesta, Spiti, Baspa, Mandakini, Dhauliganga, Trishuli and Indravati, to name a few.

In the Deccan, there are many rivers like the Krishna, Mahanadi, Tapi, Vaigai, Vegavathi, etc.

Poets sang songs about rivers and sages composed their sacred books on their banks. This immortalized the rivers, which were a source of water for irrigation and consumption.

This was the ancient Indian way of ensuring that the rivers were respected and remained free of pollution. Unfortunately, this sanctity is limited to rituals today as raw sewage and toxic chemicals are pumped into them, converting them into a source of disease and doom.

Teertha

According to the *Rig Veda* (X.40.13; X.114.7), teertha means a passage or ford in a river. Teerthas are places made sacred by the presence of a great saint or some peculiar characteristic. The *Skanda Purana* says that a place on earth resorted to by ancient good men for the collection of merit is called a teertha and the main objective is to see those holy men, although pilgrimage is only a secondary objective (1–2.13.5.10). It is a place of religious or meritorious acts; almost all virtues and acts of goodness are called teerthas.

The *Rig Veda* suggests that teertha means 'a way or road' (I.169.6) and (IV.29.3) or 'a ford in the river' (VIII.47.11). Yet, in other cases, it refers to any holy place, generally by the sea, river or an expanse of water made sacred by important events or the presence of great sages.[23] Places such as Kashi are pilgrimage sites.

The *Brahma Purana* classifies teerthas into four categories: *daiva* created by the gods (such as rivers, the towns of Kashi, Pushkar, Prayag, etc.); *asura*, created by the demons (such as Gayasura); *arsha*, created by rishis (such as Badari, Prabhasa, etc.); and *manusha*, created by men like Puru, Manu, etc. The Vedic sacrificial cult was transformed into teerthas, ensuring its spread along the banks of waterbodies. Sacrifices were conducted on the banks. Though the legends associated with the teerthas in the Puranas glorify their efficacy, they mention that merely a holy bath is not enough. The real pantheistic nature of the Hindu religion is reflected in these tales. The quality of *dana* at the teertha is fused with the social religion of the Puranas.[24]

Waterbodies

Lakes and tanks are an integral part of India's highly evolved traditional water management systems. In areas like the Deccan peninsula, where the rivers are neither snow-fed nor perennial, the different kinds of tanks—percolation ponds, natural lakes, artificial reservoirs and temple tanks—have been of great use. While the ponds, lakes and artificial reservoirs were used for activities like irrigation and washing, the temple tanks were sanctified and the waters were drawn only in times of drought.

Tanks and wells have always been the chief source of irrigation in India, maintaining groundwater levels and providing drinking water. All around India people have evolved systems to meet their water requirements. The general wisdom was to build more tanks, ponds or wells than needed, so that there was surplus water. These structures acted as flood and drought-control mechanisms. They caught water that would otherwise flood villages or make rivers to overflow. Water percolated into the soil and replenished the groundwater, nature's great reservoir.[25] To ensure that the waterbodies were maintained, they were regarded as sacred and the act of desilting and maintaining them was both an act of piety and an act of noblesse oblige by rich landlords and rulers. In the Deccan, the clay-based silt was used to make the terracotta images of Ganesha, to be worshipped during Ganesh Chaturthi, which coincided with the south-west monsoon. Till the twentieth century, the images of Ganesha were never baked. They were given red eyes from the seeds of the Indian coral tree. After three days of the puja, the image of Ganesha was put back into the waterbody, to become silt again and line the base of the tank till next year.

Lake Manasarovar (Tso Mapham—'The Undefeated'— in Tibetan, in honour of Milarepa's success in his encounter with Naro Bon-Chung) in Tibet, created from the mind of Brahma, the Creator, is the most sacred. According to a legend, Brahma's sons, who were rishis, went to Kailas and remained there in meditation for many years. Lacking a convenient place to perform their ablutions, they had appealed to their father, who created the lake from his mind (*manas* means mind, *sarovar* means lake). Then a great linga arose from its midst, which the Saptarishis saw and worshipped.

According to another legend, the Naga king and his subjects live in the lake and feed on the fruit of a giant *jambu* tree that grows in the middle. Some of the fruits fall into the water and sink to the bottom, where they are transformed into gold. Interestingly, gold has been found near the north-west corner of Manasarovar.

Lake Manasarovar is a beautiful wetland, located 4500 m high in the Himalayas. The waters come from the melting snow of Mount Kailas, which first flow as River Karnali and then into the lake. It has little vegetation—some grass—and much grey dirt, not to mention the triangular sacred stones along the lakeside venerated by Hindus as mini lingas. The blue of the fresh water contrasts strikingly with the white of Mount Kailas. Moonrise over the lake is spellbinding, with the Saptarishis (stars) shining brightly above and in the water. There is a wealth of bird life, and the dogs (Tibetan mastiffs) on the banks often try to catch them (unsuccessfully). There is one type of bird—white with gold-and-brown feathers (a waterbird)—which warns the others when a dog tries to catch it. It is known as Uma because it makes a sound like 'uma'

when warning of danger. There are black and white swans on
the lake. It is a scene of incredible beauty and romance. The
lake's beauty has been lavishly praised in Kalidasa's *Meghaduta*,
in particular the rich blue of its waters, which range from
a light blue near the shores to a deep emerald green in the
centre. It is an impressive foreground to the silver dome of
Kailas shining away in the north.

Lake Manasarovar has both religious and mythological
associations. It is the largest freshwater body in the world.
Earlier, eight gompas surrounded its margins, representing the
Buddhist Wheel of Life—with the hub at the centre and the
eight gompas situated at the points where the eight spokes
intersect with the rim. A complete circumambulation of the
lake, passing by way of the eight gompas, is considered to be
a single symbolic turn of the wheel, with all the benefits that
this implied. The length of a parikrama is generally 100 km;
a circuit of the lake along its shores is about 85 km, but this
may only be performed in winter when the streams are frozen.
Hindus, Buddhists, Jains and Bons perform the parikrama of
the lake (Figure 3.6).

Figure 3.6: Lake Manasarovar

On the other hand, Rakshastal (to the west of Manasarovar), literally 'lake of the demon' in Sanskrit, is the place where Ravana, the demon king of Lanka, performed a severe penance. According to one story, he made a daily offering of one of his ten heads on an island in this lake as a sacrifice to please Shiva till, on the tenth day, Shiva granted Ravana the superior powers he wished for. According to another story, Ravana bathed in this lake before trying to lift Mount Kailas on which sat Lord Shiva and Parvati. Shiva pressed the mountain down with his toe and pushed Ravana into the ground. In desperation, Ravana composed and recited the *Shivastotra* in praise of Shiva, after which he was permitted to live. According to yet another theory, Rakshastal was created by Ravana in order to obtain superior powers through his acts of penance to Shiva who resided on Mount Kailas.

Rakshastal is a strange contrast to the sacred Lake Manasarovar. While the latter is round like the sun, the former is shaped like a crescent. The two are regarded as 'brightness' and 'darkness', respectively. Manasarovar is situated about 15 m higher than Rakshastal, from which it is separated by a narrow strip of land and a channel taking fresh water from the former to the latter. There is an atmosphere of depression haunting the Rakshastal, and one avoids prolonging one's stay there, travelling quickly over hills and ridges to reach Manasarovar with its turquoise waters, which provides a relief from the desolation prevailing around the former. Its salt water, in stark contrast to the fresh water of Lake Manasarovar, means that there are neither aquatic plants nor any fish. The lake is considered poisonous by locals. There is a belief that the short channel Ganga Chhu which connects the two was created by rishis to add pure water from Manasarovar. There are four islands in Rakshastal which are used by local people during winter for pasturing their yaks (Figure 3.7).

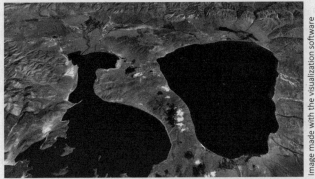

Image made with the visualization software NASA World Wind

Figure 3.7: Satellite view of Lake Manasarovar (right) and Lake Rakshastal (left)

Kurukshetra, the site of the great Mahabharata war, abounds in water reservoirs. The Bhishma Kund is located in Narkatari village near Kurukshetra. The reservoir was built long ago but was recently renovated. According to the epic, during the great battle of Mahabharata, Bhishma was defeated by Arjuna's arrows. Later, when the Kauravas and Pandavas came to pay homage to their grandfather, Bhishma asked for water. Arjuna is said to have shot an arrow into the earth, and water gushed forth to quench his thirst, which is the kund. There is a small temple near the kund that commemorates this event. Pilgrims come here from all over the country to pay homage to the great Bhishma Pitamah. It is believed that those who take a dip in this kund are cleansed of their sins and receive the strength to walk the path of righteousness.

Ararak Teertha in Narkatari village is a famous sacred tank visited by many pilgrims. It is said that this teertha is surrounded by Lord Shiva in the south, Lord Vishnu in the north, Lord Brahma in the east and the Mother Goddess in the west.

Brahma Sarovar is an ancient sacred water tank in Thanesar in Haryana. According to legend, Lord Brahma created the universe from Kurukshetra, after a huge yajna. The sarovar is also described in the *Kitab-ul-Hind* by Al Beruni in the eleventh century CE. It is believed that Duryodhana, the eldest and last of the Kauravas, hid himself beneath the waters of this tank on the last day of the Kurukshetra war. There is a temple to Lord Shiva within the tank, which can be accessed by a small bridge.

Temple tanks are artificial reservoirs generally built as part of the temple complex. They are called *pushkarni, kalyani,* kund, sarovar, *baoli/baori, vav, teertha, talab, kovil kulam,* etc. in different languages and regions of India. Many temple tanks are said to cure various diseases and maladies. This is probably due to the fact that the *abhishekam** water, filled with turmeric and herbs, is poured into the tank. It is possible that the Great Bath of Mohenjo-Daro in the Indus–Sarasvati region was a temple tank, probably the oldest in the world. Pilgrims are permitted to wash their faces and feet in these tanks. Most temples have a separate tank for washing and another for the abhishekam ritual.

The temple tanks are revered no less than the temple itself. Their waters are believed to be as sacred as the Ganga with the ability to cleanse all sins. In fact, devotees are required to wash their hands and feet in the tank before entering the temple. The waters are also used to perform the daily ritual bath of the temple deity. Annual float festivals are conducted in the tanks, when the idol of the deity is floated around the tank on

---

* A religious ritual of bathing the deity.

a decorated raft. Since the water of the tanks is not extracted for everyday chores, it has served the vital role of recharging underground aquifers, reducing water run-off and enhancing the water stagnation time, thus ensuring sufficient water in the domestic wells during the summer months. They also add aesthetic value to the temple. In small villages and towns, the temple tanks and their steps serve as a meeting place for the entire community, while in states like Gujarat and Rajasthan, the 'step wells' are beautifully decorated with sculptures.

Since ancient times, the design of water storage has been important in India's temple architecture. Temple tanks—especially of western and southern India—became an art form in itself. An example of a beautifully designed and artistic temple tank is the large, geometrically designed stepped tank at Hampi, which was the capital of the famed Vijayanagara Empire. It was filled by an aqueduct and has no drain.

A stepwell is called a vav in western India and baoli/baori in north India. They were built by kings and noblemen and were richly decorated with sculptures. It is a deep masonry well with steps going down to the water level, sometimes deep underground. It was a place for aesthetic appreciation and a haven in the hot summer months.

The Chand Baori in Abhaneri village is one of the oldest step wells in the state of Rajasthan and is believed to be one of the largest. There are 3500 tapered steps encircling the well on three sides while the fourth side has a group of pavilions that are constructed one on top of the other. The famous Chand Baori at Abhaneri was featured in a movie called *The Fall* and also made a short appearance in a popular film called *The Dark Knight Rises*.

Bundi is known for its baoris or stepwells. Constructed by royalty and affluent members of society, they served as water reservoirs when there was water scarcity. The finest example is the Raniji Ki Vav which was built by Rani Nathavati-ji in 1699 CE. Adorned with finely sculpted pillars and arches, it is a multistorey structure with a shrine for worship on each floor. The steps built into the sides of the well made water accessible even when it was at a very low level.

Pushkar is an artificial lake created in the twelfth century CE and covers an area of about 5 sq. km. About 400 temples and shrines and fifty-two ghats are located near this lake. The Ramayana and the Mahabharata contain references to this lake as the Adi Teertha, which means it was much older. Legend has it that Lord Brahma created this lake and Lord Vishnu came here in the form of the boar Varaha. Guru Gobind Singh recited the *Guru Granth Sahib* on its banks and the ashes of Mahatma Gandhi were immersed here. On Kartik Purnima (full moon in the month of Kartik), lakhs of pilgrims come here to bathe in the holy waters of the lake as it is believed to cure skin diseases. As Lakshmana is believed to have performed penance here, there is a temple dedicated to him on the banks of the lake. There is also a star-shaped gurdwara called Sri Hemkund Sahib which is an important pilgrimage site for the Sikhs.

The holy tank surrounding the Golden Temple (Shri Harmandir Sahib) at Amritsar is called Amrit Sarovar. In fact, the city of Amritsar derives its name from the sacred tank. Pilgrims bathe in the tank, and drink its waters. A holy dip in the tank is a part of the spiritual baptismal ritual for the Sikhs.[26]

Ram Teerth Sarovar, located outside Amritsar, is believed to have been dug by Hanuman. The circumference of the tank is about 3 km and there are many temples on its banks. A majority of the pilgrims consider it auspicious to take a dip in the sacred tank in the early hours of the Purnamashi (full moon) night. Ram Teerth is an ancient pilgrimage centre associated with the Ramayana. It is said that Rishi Chyavana cured Maharshi Valmiki of leprosy by making him bathe in the sarovar. Ram Teerth Mela is celebrated every year a fortnight after Diwali, when pilgrims in large numbers take a holy dip in the sacred tank. After the dip, they circumambulate the tank. Women light lamps, called *tulla*, made of kneaded flour, place them on leaf plates and float them in the tank. It is believed that this ceremony absolves one of sins and is known as *tulla toarna* (the floating of tullas). Throughout the ceremony, devotional songs and hymns are recited.[27]

All the lakes were rich in biodiversity, especially aquatic life. However, in recent times, due to developmental activities, tourism, annual fairs, immersion of ashes, etc., the lakes have become polluted.

North-east India is full of sacred lakes and tanks. The Parashuram Kund, which is associated with the legend of Parashurama, is situated in north Tezu in the Lohit district of Arunachal Pradesh. According to legend, Parashurama's mother, Renuka, had gone to fetch water from the river where she looked at King Chitranathan with desire. Her husband, Sage Jamadagani, became enraged and ordered his son to kill his mother. Obeying his father, Parashurama beheaded his mother. This was a great sin and the only way he could atone for it was to take a bath in the holy Brahma Kund. Accordingly, Parashurama cut a passage from the kund for the water to come out and the place where his axe fell came to be known as Parashuram Kund. Where Parashurama had cut open a passage through the hills for the Brahmaputra to flow is Brahma Kund, situated nearby.[28] Renuka too is a goddess in many parts of India. In Himachal Pradesh, there is a

Renuka Teertha or Lake which is sacred as it is the birthplace of Parashurama, the sixth incarnation of Lord Vishnu. In south India, there are several temples dedicated to Renuka Devi.

The practice of constructing temple tanks in Assam dates back to the medieval period when the Ahom kings constructed/excavated temple tanks for the benefit of their subjects. Among these, the four largest were the Joysagar tank (318 acres), Sibsagar tank (129 acres), Gowrisagar tank (150 acres) and Rudrasagar tank, which are still extant. Half of the Joysagar tank stands underwater. It was built by the Ahom king Rudra Singha in 1697 CE in memory of his mother, Joymoti, at Rangpur in Sibsagar district. Many temples are situated on the banks of the lake.[29]

An important pilgrimage centre for Manipuris is Kangla, whose presiding deity is Lord Pakhangba. The holy tank situated here is believed to be the abode of Lord Pakhangba. There are also sacred pools like Chingkhei Nungjeng, Manung Nungjeng and Lai Pukhri.

A temple to Radha, Kunja and Kali was constructed at Ningthem Pukhri and daily worship is performed according to Ramandi traditions. A statue of Hanuman is worshipped in a temple at Mongbahanba. A square pond or *pukhri*, 100 ft by 100 ft, is situated at Wangkhei Leikai.[30]

In Meghalaya, many lakes are considered to be sacred by local communities. Nartiang is one such place of religious importance. There are two temples in Nartiang: one dedicated to Durga and another to Shiva. There are also two lakes dug by Sajar Nangli—Umtisong and Myngkoi Tok. Thadlaskein and Umhang Lakes, surrounded by thick forests, are protected areas. Umhang became important because the Jaintia chief Sajar Nangli drank water from a spring here and was so enamoured by the beauty of the place that he decided to create an artificial lake, which become sacred to the people of Wataw. Thadlaskein, another lake dug by the Jaintia chief, is situated in the Jaintia Hills. According to legend, it was dug within a day using bows and arrows. The lake is sacred to local people who perform sacrifices here.[31]

Sikkim is home to several sacred lakes, including the source of the River Teesta and Rangit. They are also regarded as the abode of local guardian spirits such as devas, yakshas, nagas, apsaras, demons and various Tantric deities. Some of the important sacred lakes are Gurudongmar and Tsho Lhamo in north Sikkim; Tsomgo in east Sikkim; and Khachoedpalri (Khechopalri), Kathok and Yuksam in west Sikkim. Gurudongmar in north Sikkim is one of the 109 *tshochen* or major lakes of Sikkim as per the text *Nay-sol*. The lake was first recognized by Guru Padmasambhava, popularly known in Sikkim as Guru Rinpoche, and is the gateway to the northern territory, otherwise known as the hidden land of Demojong. Today, the lake has become one of the important pilgrimage sites of northern Sikkim. In Sikkim, the holy lakes are still revered as the abodes of the devas, nagas and yakshas. Local people believe that whoever worships here with devotion will definitely have their prayers answered. It is also popularly believed that Guru Nanak himself rested on the banks of this lake on his way to Tibet. At the Lachen Gompa, the guru's footprints, his robe and water utensil are still revered.[32]

Tsomgo (Changu) is a glacial lake situated 40 km from Gangtok in Sikkim. In the Bhutia language, 'tsomgo' means source of the lake and '*changu*' means above the lake. The name is believed to have been given by yak herders when, in an earlier time, they lived just above the lake. The lake, about 1 km long and 15 m deep, is situated at an altitude of 3780 m and remains frozen during winter. It is an important pilgrimage

site where the deities are worshipped in small gompas near
the lake and prayer flags hoisted around it. There is a small
Shiva temple on the bank of the lake. In the past, lamas used
to study the colour of the water of the lake to forecast the
future. It was said that if the waters of the Tsomgo Lake had
a dark tinge, it foreshadowed a year of trouble and unrest.
The star-shaped gurdwara, Sri Hemkund Sahib, located on
the banks of the lake is an important pilgrimage site for Sikhs.
For Hindus, Hemkund is where Lakshmana did penance. The
mythological name for Hemkund is Lakpal. There is a temple
dedicated to Lakshmana on the bank of the lake. The river
flowing through this valley from Gobindghat to Gobinddham
is called Lakshman Ganga. The lake is woven with rich legends
and folklore and is believed to have been originally situated in
a different location named Laten, several kilometres away from
its present location. A story goes that an old woman at Tsomgo
dreamt one night that the lake at Laten was to shift to Tsomgo.
She and her herder friends were warned to leave the place.
The woman told her friends about the dream and the warning,
but no one listened. She milked her yak and poured the milk
on the ground as it was considered an auspicious gesture and
then left the place. As she was leaving, she saw an old lady
with strikingly long white hair and fair complexion carrying
yarn entering Tsomgo. It is believed that this lady with white
hair was actually a water nymph. The coming of the lake to
the region was considered a good omen. It is worshipped as
a deity, with local people praying to it for their well-being.
Ceremonies are held during Guru Purnima.[33]

Khecheopalri Lake, situated at an altitude of 6000 ft, is
considered to be a wish-fulfilling lake. It is known as Tsho–Tsho
in Sikkimese. The lake is surrounded by dense forests. There is
a legend that it was once simply a grazing ground for cattle. The

local Lepchas used to collect nettles to make various articles. One day they saw a pair of conch shells flying towards them, which went into the ground, giving rise to an enormous spring. The lake is considered to be sacred by both the Buddhists and Hindus of Sikkim. As it resembles a footprint, local people believe it is Lord Shiva's footprint. It is believed that if even a single leaf falls on the lake's waters, it is immediately picked up by a bird. Later Buddhist saints named it Khachoedpalri (mountain of blissful heaven) and recognized it as an abode of Tsho–sMan Pemachen, the protective nymph of Buddhism.[34]

Tripura is famed for its many temples and places of religious worship. In its capital Agartala, there are many temples dedicated to various gods and goddesses, which were constructed when Tripura was ruled by kings.[35] Kamala Sagar Lake is an important waterbody, a huge artificial lake situated about 27 km from Agartala and constructed by Maharaja Dhanya Manikya Bahadur in the late fifteenth century. The sacred pond Kamala Sagar (lotus pond) was named after Maharani Kamala Devi, wife of Maharaja Narendra Manikya.[36] Dumboor Lake is another sacred waterbody, with an area of 41 sq. km and thick forest all around. There are forty-eight islands situated in the middle of the lake which attracts migratory birds. The lake originates from River Gomati at a place called Teerthamukh and is the confluence of the Rivers Raima and Sarma. The lake also contains a rich reservoir of natural and cultured fish.[37]

Kali Pokhri is a sacred pond in Darjeeling situated at an altitude of 3108 m above sea level. The dark water of this pond does not freeze in winter.[38]

Bindusagar, one of the most important sacred tanks of India, is situated near Lingaraj Temple in Bhubaneswar, Odisha. Bindusagar means 'the water body with fine suspension of silt'.

The silt makes it shine brilliantly in the sunrays, imparting a grey colour to the ripple of the water. It is believed that this place was once known as Ekamra Vana. After destroying the asuras, Parvati, who was thirsty, appealed to Lord Shiva for water. The Lord created this sagar to quench her thirst. It is believed to be fed with waters from all the holy rivers of India. A whole host of temples is situated around it. Many Hindu scriptures, such as the *Padma Purana*, *Shiva Purana*, *Brahmanda Purana*, *Kapila Samhita* and *Ekamra Purana*, have praised the religious merit of the water in the Bindusagar. A dip in it before entering into the temple is considered auspicious. However, of late, it has become polluted, with sewage water and run-off from the nearby paddy fields entering the tank. It also serves as a dumping site for leftovers from the nearby Ananta Vasudev Temple. It is also used as a bathing ghat, a place for washing clothes and for performing ceremonies. It is especially sacred because of its association with Shri Kapil Dev, the founder of Samkhya philosophy, who preached the way to attain moksha to his mother at this site.[39]

Narendra Sarovar at Puri is one of the most sacred tanks of Odisha, covering an area of 3.24 hectares and situated in a picturesque surrounding about 2 km north-east of the famous Puri Jagannath Temple. There is a small temple on an island in the sarovar dedicated to Lord Jagannath, Balarama and Subhadra. During the Chandana Yatra, the boat festival is held in the tank. Lord Madanamohana (an incarnation of Lord Jagannath) goes for a boat ride on a decorated float, locally known as *chapa*. This tank is also called the Chandana Pushkarani after the famous yatra of the same name. According to legend, Narendra Deva, the brother of Gajapati Kapilendra Deva, excavated this tank and so it was named after him. Gajapati Kapilendra Deva also constructed sixteen ghats named after Narendra Deva, his wife and their fourteen sons. The Chandana Bije Ghat (Lamba Chakada) was constructed for the purpose of conducting the *chandana bije* of Lord Jagannath. This *chakada* is also named after Narendra Deva, who also constructed a temple of Kalandishvara Shiva and Gopinath on the bank of the holy tank. Brahma Jaga, named after the court poet Narahari Brahma, was also established on the eastern side of the Narendra tank.

There is also a myth that the creation of the Narendra Sarovar is due to a pumpkin seed. King Narendra Deva was a great devotee of Lord Jagannath. Once he found a pumpkin seed in the palace courtyard and gave the seed to his *sarbarakara* (revenue collector) to be planted in the name of the Lord and ordered that all the pumpkins be offered to the Lord. The creeper produced lakhs of pumpkins, which the sarbarakara sold. The money earned was handed over to the king, who then offered it to the Gajapati king of Puri. Both the kings decided to construct a tank at Shrikshetra out of this money. Thus, the famous Narendra tank covering an area of 14 acres came into existence.[40]

South India, being rain-fed, is heavily dependant on temple tanks for maintaining groundwater levels. Every temple has one or several tanks, with steps going down into the water. Haridra Nadhi, the temple tank of the Rajagopalaswamy Temple in Mannargudi (Tiruvarur district in Tamil Nadu), is the largest temple tank in India. It covers an area of 23 acres (93,000 sq. m). It is also known as the daughter of the Kaveri.

The kalyani, pushkarni, kovil kulam, etc. are ancient Hindu stepped tanks used by pilgrims for bathing before entering the temple. These tanks were built near temples to enable the pilgrims to bathe and clean themselves before prayer. During summer, they were desilted and the clay was used to make images of Ganesha—unbaked—for the festival of Ganesh Chaturthi. After the festival, the unbaked images were immersed in the water tank, as a result of which it melted. Nowadays, they are still used for the immersion of Ganesha idols, but as the idols are baked and painted, they pollute the waters and do not melt away.

In Tamil Nadu, Karnataka, Andhra Pradesh and Kerala, every village has a temple and a tank and cities have many temples with many tanks. There are many sacred tanks too—one or more for each temple.

The most famous of the tanks in Tamil Nadu is the Mahamakam tank in Kumbakonam. According to legend, at the end of each era, the whole world is immersed in a deluge by Shiva for human sins and life on earth is recreated by Brahma. At the end of the previous *yuga*, Shiva sent a divine pot. When the divine pot reached Kumbakonam, Shiva, as a hunter, broke it with an arrow. The pot was broken and fell far and wide, becoming the origin of many temples—Kumbeswara, Someswara, Kasi Viswanatha, Nageswara, Kamata Viswanatha, Banapuriswara, Lakshminaryana, Sarangapani, Chakrapani and Varadaraja. Brahma prayed to Shiva to permit pilgrims to visit the tanks during sacred occasions, and Shiva agreed. Astronomically, when the planet Jupiter passes over Leo on the day of the festival, it is believed to bring all waterbodies together and make them as sacred as the Ganga. Millions bathe in the tank on every Masimaham, an annual event that occurs in the Tamil month of Masi (February–March), under the Makam star. Once in twelve years, when the planet Guru (Jupiter) enters Simha (Leo), the festival of Kumbh Mela is celebrated here.

The Varadaraja Perumal Temple at Kanchipuram has several sacred tanks, some of which are located outside the temple. The main temple tank is called Ananta Saras Pushkarni. The ancient wooden image of Varadaraja made from a fig tree—known as Atti Varadar—is preserved in a silver box in the tank, from which water is pumped out every forty years.[41] The murals on the walls of the temple depict the 108 *divya desam*s (temples of Lord Vishnu) and their tanks. Many temples have sacred tanks in a variety of shapes—square, round, rectangular—and even rivers flowing by them.[42]

The Swami Pushkarni on Tirumala Hill, in front of the temple of Balaji, is believed to have been Vishnu's tank in Vaikuntha, which was brought to earth by his vahana, Garuda. It is regarded as the equivalent of the River Ganga in sacredness: a dip in the tank cleanses the bather of all sins; the performance of any rite on its banks is said to cure deformities of the body and avoid all types of hell. It grants the wishes of those who bathe in it. There are seventeen teerthas in Tirumala whose waters are believed to mix with the Swami Pushkarni[43] (Figure 3.8).

Photograph by M. Amirthhalingam

Figure 3.8: Tirumala temple tank

Shravanabelagola in Karnataka has a beautiful tank overlooking a hill. Shravanabelagola means 'white pond of the Shravana' and refers to the colossal image of Gommateshvara. At the entrance to the complex is the Manasthambha, a highly decorated pillar with a small-seated figure of Brahma on top, facing east. The prefix 'Shravana' serves to distinguish it from other *belagolas* or 'white ponds', an allusion to the enormous sacred body of water in the middle of the town. Shravanabelagola has two hills: Chandragiri, named after Emperor Chandragupta Maurya, who is believed to have renounced his throne, become a Jain ascetic and moved south with his guru Acharya Bhadrabahu or Bahubali; and Vindhyagiri.

The temple of Padmanabhaswamy in Thiruvananthapuram, Kerala, stands beside a sacred tank, the Padma Teertham (lotus waters).

The waters of the sacred tanks are believed to cleanse all sins. In fact, devotees are required to wash their hands and feet in the temple tank before entering the temple. But the holy water bath does not distinguish between the sins, be they heinous crimes or misdemeanours, and mostly bestows one's original form. Some teerthas are capable of blessing a person with a particular form or a better form depending on their powers.[44]

Apart from performing the abhishekam of the temple deity, annual float festivals are conducted in the tanks, wherein the idol of the deity is floated around on a decorated raft.

The temple tanks were in fact rainwater harvesting structures which served the vital purpose of recharging underground aquifers. They reduced the run-off and enhanced the water stagnation time, which ensured sufficient water in the domestic wells during the summer months. Some of these sacred tanks supported a variety of life forms, especially fish, which helped maintain the tank by eating moss and algae, which would otherwise turn the water murky. In fact, fish were actively bred in many temple tanks for their maintenance. Most of the ancient tanks have fallen into a state of disrepair and disuse because of unchecked extraction, and blocking of inlet ducts has led to the drying up of many. Pressures on land have led to the encroachment of many dry tanks. For example, in Bengaluru, the famous Dharmambudhi tank has been drained to make way for the Majestic bus stand. The tanks have also become sinks for the sewage and garbage of the neighbourhood. Those that still have water have been invaded by various kinds of weeds, rendering them unfit for use.[45]

Figure 3.9: Kandaswami temple tank, Chennai.

Chennai city used to have about 250 small and big waterbodies in and around it, but today, the number has been reduced to twenty-seven. A 300-year-old map of Madras showed about 250 reservoirs, over seventy temple tanks and three freshwater rivers which are now highly polluted and unusable.[46] The Kandaswami Temple tank in Georgetown, Chennai, is one of the few well-maintained tanks in the city (Figure 3.9).

4

Plants as Protectors[1]

A person is honoured in Vaikuntha for as many thousand years
as the days he resides in a house where the *tulasi* grows.

And if one grows the *bilva* properly, which pleases Lord
Shiva, the goddess of riches resides permanently . . .

He who plants even a single *ashvattha*, wherever it may be,
goes to the abode of Hari.

He who has planted *dhatri* has . . . donated the earth. He
would be considered a celibate forever.

He who plants a couple of *banyan* trees . . . will go to the abode
of Shiva and many heavenly nymphs will attend upon him.

After planting *neem* trees a person well-versed in *dharma*
attains the abode of the Sun. Indeed! He resides there for
a long period.

By planting four *plaksha* trees a person undoubtedly obtains
the fruits of the *rajasuya* sacrifice.

He who plants five or six mango trees attains the abode of
Garuda and lives happily forever, like the gods.

One should plant seven *palasha* trees or even one. One
attains the abode of Brahma and enjoys the company of
gods by doing so.

He who plants eight *udumbara* trees or even prompts
someone to plant them, rejoices in the lunar world.

He who has planted *madhuka* and propitiated Parvati, becomes free from disease . . .

If one plants *kshirini*, *dadimi*, *rambha*, *priyala* and *panasa*, one experiences no affliction for seven births.

He who has knowingly or unknowingly planted *ambu* is respected as a recluse even while staying in the house.

By planting all kinds of other trees, useful for fruits and flowers, a person gets a reward of a thousand cows adorned with jewels.

By planting one *ashvattha*, one *pichumanda*, one *nyagrodha*, ten tamarind trees, the group of three—*kapittha*, *bilva*, *amalaka*, and five mango trees—one never visits hell.

—*Vrukshayurveda* (9–23)[2]

The worship of the tree is an ancient phenomenon in Hinduism, and all over the world. Plants were sanctified for the socio-economic and health concerns of ancient people. The earliest form of worship was probably the veneration of the tree. When people became food producers, the Mother Goddess or the Earth Mother became the chief deity and plants became her blessings to her devotees. People worshipped the tree as a symbol of fertility essential for the survival of early man. Spirits—good and bad—were believed to reside in them. Trees were worshipped to please the resident spirits. When sacred forests were cleared for agriculture, a single tree was left, which was designated as the sacred tree.

The earliest temples used to be images placed under trees. Later, the tree and image were enclosed by a fence made of wood, followed even later by stone. The temple building was a later construction. Numerous references have been made in literature to trees as abodes of gods. They sheltered the object of worship: a fetish, a weapon, a deity or any other. As the open-air shrine beneath the tree was replaced by a shrine or temple for the deity, the tree became the *sthala vriksha* of the temple; the tree was associated with the deity and became an inseparable part of the local mythology. The sthala vrikshas of India constitute the single genetic resource for the conservation of species diversity. They once played a major role in local ecology and their worship celebrates our biological heritage.

But trees were recognized as being animate and having life, with the ability to feel happiness and sorrow, long before Professor J.C. Bose proved that plants have life or invented the crescograph. It is said in the Mahabharata: 'Trees take water from the roots. If they have any disease it is cured by sprinkling of medicines. It shows they have *rasendriya*. Trees are alive and they have life like others because on cutting they feel sorrow. Similarly they feel happiness. After cutting, a new branch comes out' (XVIII.15–17).[3]

The earliest instance of tree worship appears in the Indus– Sarasvati civilization. Trees were very important during this period, for so many trees appear on the seals. Some were placed on a pedestal and/or fenced in, like the sthala vrikshas of Indian temples.

The Vedas refer to the 'cosmic tree' (also called the 'tree of life'), which embraces the whole universe, the mythological 'Axis Mundi' of the Old World. For the Vedic people, the cosmic tree is the symbolic power embracing the entire universe. This tree is seen as the Goddess of Nature, who nourishes all life and is rooted in Brahman, the Ultimate.[4]

The most sacred tree of India is the pipal or *ashvattha*, whose importance goes back to the Indus–Sarasvati culture. On a steatite seal from Mohenjo-Daro, a figure with a horned headdress, long braid and bangles on both arms stands within a pipal tree. Is he the spirit of the tree, a yaksha? Apparently, the spirit of the tree, the yaksha or *yakshi*, is as old as the Indus–Sarasvati civilization, giving sanctity to the tree within which he or she dwells. On one side, there is a kneeling worshipper and a gigantic ram, possibly a sacrificial offering. Seven male

Source: The C.P. Ramaswami Aiyar Foundation's photo archives

Figure 4.1: Worship of pipal tree, Indus Valley seal

figures stand in a row below, possibly the Saptarishis of Vedic religion, identified with the seven stars of the Big Dipper or Ursa Major—a hierarchy working under the guidance of the Supreme Being. All the seven figures are dressed identically, with a single plumed headdress, bangles and folded dhotis (Figure 4.1). Even today, Hindus circumambulate the pipal seven times, chanting,

> *Mulato Brahma rupaya , Madhyato Vishnu rupine,*
> *Agrato Shiva rupaya, Vriksha rajaya te namah.*[5]

> (Whose root is the form of Brahma, whose middle is the form of Vishnu,
> Whose top is the form of Shiva, My salutations to you,
> O King of Trees.)

The holy tree removes all the sins earned in several hundred births, and grants prosperity.

There are many other examples of pipal worship. The tree grows out of a central hub flanked by two unicorn heads, with pictograms below; a male figure, probably the spirit of the tree or yaksha, stands within a stylized pipal tree, a recurring theme on several mass-produced moulded pieces as well as on seals; and a branch with three leaves grows out of the head of a three-faced male figure who is probably a proto-Shiva. There is a seated male figure, with a single branch bearing three pipal leaves rising from the middle of his headdress. The figure is seated in a yogic posture, with three faces, one facing front, one left, and one right. He wears an elaborate headdress of two buffalo-style horns curved outwards. This person could only be Lord Shiva or Pashupati, the three-faced Trimukha or Trimurti, incorporating the Creator, Preserver and Destroyer as he is celebrated in later art, such as in the eighth-century caves at Elephanta near Mumbai.

Hindus believe that Lord Vishnu was born under the pipal and Lord Krishna died beneath it. Also, since the Vedic period, Hindu holy men have been known to meditate sitting under this tree. Hence, it is very sacred for the followers of Hinduism.

The Mahabodhi tree, a pipal tree located at Bodhgaya, Bihar, is recorded as having been planted in 288 BCE by Emperor Ashoka. It is believed that Siddhartha Gautama, the founder of Buddhism, achieved enlightenment or bodhi under this tree, obtaining the appellation of Gautama Buddha. Lord Buddha has said: 'He who worships the pipal tree will receive the same reward as if he worshipped me in person.' The pipal is a source of knowledge for all: serpents, the keepers of divine knowledge and treasure, and the Buddha who achieved enlightenment beneath it (Figure 4.2).

It is the most sacred tree because it is the dwelling place of the Hindu trinity: Brahma, Vishnu and Shiva. It spreads its branches to bring blessings to all creation. It is worshipped by women for fertility and longevity. The tree is looked upon as an incarnation of Lord Vishnu and as an embodiment of Goddess Lakshmi. The pipal tree is worshipped by women for fertility and also longevity of their husbands. Ganesha idols or snake stones are placed at the base of the tree by devotees who want to beget children. Rituals to the Goddess Savitri are

Drawing by S. Prema, The C.P. Ramaswami Aiyar Foundation

Figure 4.2: Buddha under the pipal tree

also performed beneath the tree to have children and to avoid widowhood. The roots represent Brahma, the trunk, Vishnu and the branches, Shiva. It is the 'tree of knowledge', 'tree of life', 'tree of eternal life' and 'tree of creation'. It spreads its branches to grant blessings to all creation.

The word pipal is derived from the term *pippala*. Ashvattha is the Sanskrit name for the tree, which means 'one who does not remain the same tomorrow', or the universe itself. According to Adi Shankara, the pipal tree represents the entire cosmos. Cutting it is looked upon as a great sin, while anyone who plants it is said to receive blessings for generations to come. Describing the power and attributes of the pipal, Krishna says: 'I am the *ashvattha*, lord of trees' (*ashvattha sarva vrikshanam devarshinam cha*) (*Bhagavad Gita*, 10.26).

The *Rig Veda* (X.5.10) says that the pipal, which is a tree with 1000 branches, is symbolic of the cosmos. The tree was connected to the Maruts whose offerings were made in bowls fashioned out of pipal leaves (*Shatapatha Brahmana*, IV.3.3.6). The *Atharva Veda* (V.4.3.) refers to it as the permanent seat of the gods and reverence for the pipal was reverence for the gods themselves, as this tree grew in heaven: *'ashvatthodeva sadanah'* (*Atharva Veda*, XIX, 39.6). There are references to the tree in Panini's *Ashtadhyayi* (4.3.48), Kautilya's *Arthashastra* (1.20) and Varahamihira's *Brihatsamhita* (59.5). The *Vishnu Purana* says that just as the pipal tree is contained in a small seed, so is the whole universe contained in Brahman. The *Brahma Purana* and the *Padma Purana* relate how, once, when the demons defeated the gods, Vishnu hid in the pipal. Therefore, worshipping the tree is equivalent to worshipping Vishnu, and does not require either image or temple.

The pipal often stands alone, mounted on a wide pedestal, with an altar for devotees' offerings. The tree is regarded as sacred, possessed by a spirit or *vriksha devata*, possibly the yaksha who once resided in the tree. Seven pradakshinas (circumambulations) are done around the tree every morning. If the new moon appears on a Monday, women circumambulate the tree 108 times and light little hand-made wheat lamps. Being symbolic of fertility and protection, the tree is worshipped, particularly by women, who tie toy cradles on the branches and pray for children.

Where there is a pipal, there will also be a neem. A platform is built around them and one or more images of the snake are installed on it and worshipped. This is believed to bless the worshipper with prosperity. Sometimes a Ganesha or Hanuman idol may also be installed with the snake stones and then the place becomes a sacred prayer stop for passers-by. Weddings are performed of the neem and the pipal to ensure a good rainfall. In many parts of India, the pipal is regarded as male and is ceremoniously married to a neem tree, which is regarded as female. As these two trees usually grow side by side, a platform is built around them and the wedding is performed in summer, before the monsoon, with Brahmin priests and Vedic rituals, in the belief that it symbolizes fertility and will ensure a good rainfall and harvest. This symbolic association of the sexes is reversed in Rajasthan and Punjab, where the neem tree is considered male.

The pipal tree is sacred to both Hindus and Buddhists: for the latter, it was the tree beneath which Buddha attained enlightenment. It symbolizes the recycling of life and immortality through its ever-expanding ability to sprout new roots. The pipal is usually evergreen and lives for centuries, almost indefinitely, and is considered to be masculine because it grows as an epiphyte on other trees, especially the *shami,* considered to be feminine.[6]

The sanctity of the tree can also be traced to the Vedic ritual of kindling the sacrificial fire at religious ceremonies by rubbing two pieces of wood, one of which was the pipal. The second was the shami, known today as the khejri or *khejari.* The ceremony was called the birth of agni, or fire.[7]

The shami or khejri or Indian mesquite, the desert plant of Rajasthan, also appears on the Harappan seals, associated with a female figure and a tiger, maybe a proto-Durga. This theme is repeated on several seals and bas-relief tablets from Mohenjo-Daro, where the tiger invariably looks back at a woman—goddess or *yakshini*—seated with one leg folded beneath the other, possibly in a yogic posture, on a shami tree. This is obviously the goddess of the tree, accompanied by the tiger, since no ordinary mortal would be accompanied by a tiger. The importance of trees and forests thus goes back to India's earliest civilization.[8] While the pipal is found all over the subcontinent, the khejri is a desert plant, widely prevalent in the Thar Desert in Rajasthan, which is sacred to the Bishnois who live in the desert and graze cattle.[9] The presence of this

tree on the Indus seals suggests that the desertification of the
region had already set in.

Interestingly, in the aindu tinai or the five geographical
classifications of land, the desert or paalai is associated with
Kotravai or Durga and the *kotraan*, a desert plant of doubtful
identity. It could easily have been the shami of the *Rig Veda*,
the khejri of the later Bishnois.

The Vedas endowed the shami with the property to create
fire. According to a Rigvedic legend (X.95.1–18), Pururavas
generated the primeval fire by rubbing together two branches
of the shami and pipal trees. In later literature and tradition,
the khejri is a feminine tree, identified with the Mother
Goddess as Shamidevi, while Shiva, the Buddha, and the Jaina
Tirthankara Anantanatha—all men—sit in meditation beneath
the pipal tree.

The pipal was rubbed over the shami for kindling the
sacred fire.* Pipal was male and shami, female. Their role in
producing the sacred fire gave them a special significance.
The pipal tree symbolized the universe long before Buddha
obtained enlightenment beneath it.

Many other trees are sacred because of their associations
with gods. Most have their own mythology.

The banyan tree is regarded as sacred because it is the tree
beneath which sits Shiva, as the great teacher Dakshinamurti
(Figure 4.3). It is extremely important in religious ceremonies
and its aerial roots are symbolic of the matted hair of Shiva,
the great yogi. Savitri brought back her husband Satyavan
to life from death beneath this tree. It is worshipped by
women in Uttar Pradesh, Bihar, Gujarat and Maharashtra on

---

* Five thousand years ago, there were no matchsticks. Fire was
  produced by rubbing sticks together.

the new moon day in the
month of Jyeshtha to avoid
widowhood. The day is
known as Vat-Savitri in
honour of Savitri. When
Markandeya wanders after
the Great Deluge, he finds
a male child, who identifies
himself as Narayana, resting
on a couch made from a
single leaf of the banyan
tree or vata. The child tells
him that he is the Creator,
Preserver and Destroyer.

Figure 4.3: Dakshinamurti

Drawing by Y. Venkatesh, The C.P. Ramaswami Aiyar Foundation

Yet another tree with
sacred associations due to its appearance is Krishna's butter
cup, also known as vata. Baby Krishna is generally depicted as
sleeping within the leaf of this tree because its sides are turn
upwards in the form of a cup to hold a child. According to
legend, Krishna used to make cups from the leaves of this tree
to steal curd and butter from his mother and the cowherdesses.
Hence it is also called *makkhan katori* or butter cup.

Bengal quince or bilva is associated with Shiva and is
invariably found in Shiva temples.

Similarly, neem is associated with Devi and is found in
Devi temples. It is also a great medicinal plant, which is why
Devi is associated with many dreaded diseases like smallpox,
which are generally treated with neem.

Tulsi is a sacred plant worshipped as an incarnation of
Goddess Lakshmi and therefore associated with Krishna. Saint
Tyagaraja, the Carnatic music composer, hailed tulsi as the
'Mother of the Universe' and composed six kirtans about

her.[10] The plant is generally kept in the centre of the open courtyard in Hindu homes and women worship and water it every day (Figure 4.4). On a more practical note, tulsi prevents coughs, colds and fevers, and hence it is placed in the open courtyard.

Drawing by Y. Venkatesh, The
C.P. Ramaswami Aiyar Foundation

Brahma's religious association is with the cluster fig or *udumbara*. It is a *kalpavriksha* or a wish-fulfilling tree of life which is used in the *homa* or sacred fire.

Figure 4.4: Tulsi

There are also trees which are locally important and which may be associated with one or another deity in a particular place. For example, while the sacred tree of Shiva is the bilva, the Alexandrian laurel or *punnaga* is the sacred tree in the temple of Kapalishwarar at Mylapore, Chennai. In the past it was used for building ships to sail the seas from the ancient Pallava port of Mylapore. In Vaideeswaran Koil at Kumbakonam in Tamil Nadu, the sacred tree of Shiva is the neem. In Badrinath, it is the Indian jujube or badari, beneath which Adi Shankara meditated at Badrinath. Badari is also sacred to Vishnu, who is called Badrinath, the lord of badari, and Guru Nanak, who said; 'O God, you are an infinite tree and I am a bride under thy protection.' The tree is found in several Sikh temples.

The trifoliate leaves of the *kimshuka* or flame of the forest (*Butea monosperma*) represent the Hindu trinity of Brahma, Vishnu and Shiva, the Creator, Preserver and Destroyer, respectively. Thus, a flower of this tree is considered to be the most sacred offering. The red flowers are also offered to goddess Kali. They are believed to be the dwelling place of gandharvas and apsaras.

Similarly, the *nagapushpam* or Indian cannonball tree is sacred to Shiva. It is used in the daily worship of the lord and is given the prefix 'naga' because its petals resemble the hood of the cobra, which is sacred to Shiva.

Champaka is believed to have been brought from heaven by Krishna. Its gentle scented flowers are used in the worship of Lord Krishna and Lord Shiva. It is also regarded as an incarnation of goddess Lakshmi. According to legend, Lord Brahma promised that those who plant two champaka trees will go to heaven. It is considered to be a life-giving tree.

Mango, the most loved fruit of India, is a plant of prosperity and auspiciousness. Its leaves are used to make the decorative *toran* at the entrance of the house, while five mango leaves flank the coconut in the *kalasha* or auspicious pot. Mango leaf is used as a spoon for pouring ghee into the sacred fire. This plant is considered to be Prajapati himself, the lord of all creation. The leaves of the tree have antimicrobial properties, thus establishing the fact that the practice of decorating the home or the entrance with mango leaves was actually meant to check infection. It also acts as a collector of dust particles. The mango is the national fruit of India.

The coconut, with three eyes, is a very auspicious fruit. It is always placed on the sacred kalasha. It is gifted by worshippers to temples where it is broken in front of the deity. Coconut is an essential part of every Hindu religious ceremony and a symbol of good fortune and prosperity. It is referred to as *shriphala* or the fruit of Lakshmi, the goddess of prosperity, symbolically depicted by the poorna kumbh or 'vase of plenty', consisting of a pot filled with water or rice and topped with a coconut. In south India, it is auspicious to distribute coconuts to married women during weddings and festivals. Among Tamils, the *mangal sutra* or *thali* worn by the bride, consisting of a piece of turmeric or a gold pendant strung on a yellow thread or gold chain, is initially tied around the coconut before it is tied around the bride's neck by the groom. In Sanskrit, it is called *narikela*, which means water springs.

Amla or Indian gooseberry is worshipped as the Earth Mother because its fruit is considered to be healthy for all mankind. The other name for this plant is *dhatri*, which means nursing mother. The amla tree is worshipped on Shivaratri day, when red and yellow threads are tied around its trunk.

The lotus plant is celebrated in literature for its beauty and fragrance. It is believed that it sprang out of Vishnu's forehead, out of which came Shri, another name for Laksmi. Hence it is known as *shripushpa*. Lord Vishnu is represented with a lotus emerging from his navel, hence the name Padmanabha (lotus navel), on which sits Lord Brahma.

Both Lakshmi and Sarasvati sit on the lotus. In fact all the gods and goddesses of Hinduism are shown seated on the lotus or holding it in their hands some time or the other. According to Buddhism, the Buddha was born out of a lotus. The sanctity of the plant comes from the fact that it represents purity of body, speech and mind. This plant grows from a stem deep in the mud underwater and the flower comes out to face the sun (Figure 4.5).

Another sacred plant is the plantain which is referred to in

Drawing by Y. Venkatesh, The C.P. Ramaswami Aiyar Foundation

Figure 4.5: Goddess Lakshmi
on the lotus

the epics. The plant is dressed like a bride in many places and worshipped as Goddess Lakshmi. It is also placed in front of a bilva tree and worshipped as an incarnation of Durga. The plantain is one of the *navapatrika* (nine leaves obtained from nine sacred plants). The sanctity of the plant is derived from the fact that the banana fertilizes itself without cross-pollination. It represents fertility because it grows in clusters. Its leaves are used to place the sacred offerings in front of the deity.

Turmeric is very auspicious and used extensively in Indian ceremonies, festivals and weddings. In the weddings of many Hindu communities, a piece of dried turmeric root tied on a thread that is coloured with turmeric powder is the mangal sutra or thali that is tied around the bride's neck by the groom. Today, it has been replaced by a gold pendant. The turmeric is not only a culinary spice, but also a medicinal plant. The curcumin in turmeric has been found to prevent dementia and Alzheimer's.

Cotton-wool grass or *darbha* is sacred and is used in ceremonies, particularly those involving Ganesha, the elephant-headed god, because elephants love to eat darbha grass.

Many cities are associated and even named after plants. Vrindavan near Mathura is named after the sacred vrinda or tulsi plant. Badrinath is named after the badari tree. Kanchipuram is named after the *kanchi* tree. Goddess Meenakshi of Madurai is also known as Kadambavanavasini, meaning one who dwells in the forest of *kadamba* trees, where her temple was once situated.

All plants that are deemed to be sacred are invariably medicinal plants, used either in home remedies or in Ayurveda. The few that are not medicinal have important local economic value.

Plants are very important for locating a text. The *Rig Veda* lists about sixteen plants, including aquatic plants and grasses, all of which are found in the plains of India and modern Pakistan.

Trees, grasses and herbs were regarded as divine. Trees, says the *Rig Veda*, are the lords of the forest (Vanaspati), self-regenerating and eternal, the homes of the gods (X.97). While many people debate the origins of the Vedas, there can be no denying the authenticity of the geography they describe. People will only write about the plants and animals they are familiar with. There are no cold climate plants in the *Rig Veda*, not even the Kashmiri deodar.

The 'Papavimochana Sukta' of the *Atharva Veda* (XI.5, 1–23) has an invocation to numerous deities for deliverance from distress, which includes trees (vanaspati), herbs (*aushadhi*) and plants (*virudha*).

In the Vedic period, all of nature was divine. An indivisible life force united the world of humans, animals and plants. Besides trees, several grasses and herbs were also sacred. Trees are the homes and mansions of the gods (*Rig Veda*, X.97.4). The udumbara (cluster fig) was used for making the *yupa* (sacrificial pole), udumbara and *khadira* for making the *shruva* (ladle), *nyagrodha* for making the *chamasa* (sacrificial bowl), and bilva for making the yupa for the sacrifice. The tree is the 'lord of the forest . . . whose praises never fail' (*Rig Veda*, IX.12.7). Long before the Buddha sat beneath the pipal tree, Shiva sat beneath it in the Indus–Sarasvati region. The *Rig Veda* (X.97.5) tells the Supreme Being: 'Your abode is the *ashvattha* tree; your dwelling is made of its leaves.'

Trees are not only Vanaspati or 'lord of the forest', but are invoked as deities along with the waters and mountains. There are several references to tree worship in the Vedas. *Atharva Veda* (XIX.39.6) says the pipal is the abode of the gods; the gods sit beneath the tree in heaven. The *Atharva Veda* refers to folk cults, including those that worshipped trees, spirits and demons. Trees had mysterious abilities to grow, cure and help people in need. That is probably why Aranyani is not a mere goddess; she is the spirit of the forest (*Rig Veda*, X.146).

There are also numerous hymns and verses where the tree was not only an object of worship but also a symbol of the cosmic tree of life. The *Shvetashvatara Upanishad* (3.9) says that a tree of huge dimensions filling the space around it with its numerous branches and foliage, with a lofty trunk and many stems rising high in the sky, was the symbol of the cosmic tree or Brahman. This could only be the banyan, with its aerial roots. The *Rig Veda* mentions the word *naichhashakha*, which has been described as 'worshippers of the banyan tree'. *Atharva Veda* (V.4.3) says that it is prohibited to cut the *vata vriksha* (banyan tree) because gods live on the tree and there is no disease where this tree is situated.

According to the *Chandogya Upanishad* (VI.11), the cosmic tree symbolizes life while the *Taittiriya Brahmana* equates Brahma with the forest and the tree: '*Brahma tad vanam, Brahma sa vriksha asa*' (Brahma is the forest, Brahma is the tree) (*Rig Veda*, X.81.4; *Taittiriya Brahmana*, II.8.9.6). Several other trees are listed as sacred: pipal, banyan, shami, flame of the forest, cluster fig and Indian gooseberry (*Chandogya Upanishad*, VI.11). The Indian plum is frequently mentioned in the *Atharva Veda*. Its wood was used to make sacrificial bowls.

The *Mundaka Upanishad* describes the tree of life on which dwell two birds in eternal camaraderie, one of which eats the fruits while the other looks on silently. The two represent vitality and reflection, Nara and Narayana, the man and the divine.

Varuna is the root of the tree of life, the source of all creation (*Rig Veda*, I.24.7), a great yaksha reclining in *tapas* on the waters from whose navel springs a tree (*Atharva Veda*,

X.7.38). In the *Yajur Veda* (V.6.4), this quality is inherited by Prajapati and, in the Mahabharata, by Narayana. The tree is the pipal but later the Creator was seated on the lotus, which issued from Narayana's navel. The pipal was held in great esteem in the Vedic period. A tree of huge dimensions, it was symbolic of the cosmos and even of the Brahman. The *akshaya vata*, the eternal tree, was the basis of a profound metaphysical doctrine in Vedic literature.

There are sacred trees for each yuga.* The sandalwood tree (*Santalum album*) is the sacred tree of the *satya* or *krita yuga*, champaka (*Michelia champaca*) of the *treta yuga*, the capper bush (*Cleome fruiticosa*) of the *dwapara yuga* and the jackfruit tree of the *kali yuga*.

The Ramayana mentions various types of sacred trees: the *rathya vriksha* or roadside trees (II.3.18.50; V.12.18.22–29) and the *devtanishthana vriksha,* which were the abode of deities. The latter were divided into *yaksha chaitya* (the tree with the spirit within) and *vriksha chaitya* (the protector tree). There were also the *chaturpathavarti vriksha* (tree at the junction of crossroads with revetments around the trunk) and *shmashana vriksha* (trees grown in burning ghats). The yaksha chaitya and vriksha chaitya became, in time, the sthala vrikshas or sacred trees of temples, in memory of the trees that had once grown in plenty in that region.

There were strict injunctions against the felling of trees in Lanka. Ravana says that he had never cut down a fig tree

---

* A yuga is an era or epoch. The first was the satya or krita, which lasted for 17,28,000 human years; the next was treta, which lasted 12,96,000 years; the third was the dwapara yuga, which went on for 8,64,000 years; the present era is the kali yuga, which shall last for 4,32,000 years.

in the month of Vaishakha (April–May). Hence, he wonders why this cruel fate had befallen him. The Ramayana observes that even in the kingdom of Ravana, the planting of trees was considered a worthy practice. There was a popular belief that the cutting of trees would bring about the destruction of the woodcutter and his family.[11]

When Rama and Lakshmana were searching the forest for Sita, they came across a badari tree and asked it whether it had seen Sita. The tree answered in the affirmative and pointed in the direction in which Sita had gone. Pleased, Rama blessed the tree and gave it a boon that it would never die. In another incident, Rama came across Sabari, a poor tribal woman who was his great devotee. She tasted each and every badari fruit to see whether it was tasty before offering it to Rama. Rama said that anything offered to him with a pure heart and genuine love was clean and pure, and Sabari's soul was liberated. Since then, the fruit has been regarded as sacred and is included in religious ceremonies.

The darbha grass, cut in bunches and spread out as a seat with its tips pointing eastwards, was used for sacred purposes (Ramayana, I.3.2), while it was placed in the opposite direction during the shraddha or death ceremony (Ramayana, II.104.8). The fresh leaf blade, elongated and pointing with sapphire-like lustre, was used as a missile by Rama (Ramayana, V.38.29). The religious importance of kusha grass is reinforced after Sita enters the earth to leave this world. She prays to her mother, the earth, to take her back. The earth opens up. Her son Kusha runs forward to save her but could grasp only her hair, which turns to grass. Named after Kusha who tried to save her, the grass is held sacred and used in various rituals (Ramayana, 'Uttara Kanda').

The Mahabharata (III.187–89) also says that during the maha pralaya (great flood), the sage Markandeya was wandering

around the great abyss of water when he saw a child floating
on a leaf of the banyan tree. The child identified himself as
Narayana, the resting place of souls, the Creator and Destroyer
of the Universe. Later, as the cults of Narayana and Krishna
coalesced, the child on the leaf became Krishna, the Vata-
patra-shayi of later art. In the Tamil *Perumpanatruppadai*,
King Ilan Tiraiyan belonged to a tribe that 'descended from
the waves', whose ancestor floated down the sea tied to a
creeper (*tondai*),[12] after which Tondaimandalam, the Madras–
Kanchipuram region, was named.

Krishna furthered the worship of trees. He lived in a
forest and told his friends to worship the beautiful trees that
lived for the benefit of others. Every part of the tree, he said,
is useful and anyone who approaches a tree benefits from
it. Krishna advised the cowherds to follow the example of
the tree in their lives. His attribute was the flute, a simple
wind instrument made of bamboo, which he played in the
vrinda or tulsi forest on the banks of the River Yamuna on
moonlit nights. His favourite tree was the kadamba. Krishna
fought Indra to bring the divine *parijata* (night-flowering
jasmine) tree to his new city of Dwarka. His love of trees was
translated into *Nikunja Lila* (play in the bower or arbour) by
the Premopasaka Sampradayis of Vraj, according to which
devotees should meditate upon the divine couple Radha and
Krishna's nikunja lila and attain spiritual bliss as the path for
attaining total happiness.

The Mahabharata (Southern Recension, XII.69.41 ff)
says that holy trees should not be injured as they are the abodes
of devas, yakshas, rakshasas and so on. The epic says that the
yaksha is a *vrikshavasin* (tree-dweller) (II.10.399) and that trees
have life (II.18.15–17). It has several references to the worship
of plants. Many were regarded as deities. Plants such as the

basil, pipal, banyan and Indian gooseberry were worshipped
by the common people. In the 'Adiparva' (138.25), the epic
says that the human heart makes man regard plants and trees
with sanctity and as worthy of worship.

Yakshas and Yakshis

The *Atharva Veda* (X.7.38) describes 'a great *yaksha* in the
midst of the universe reclining in concentration (*tapas*) on the
back of the waters, therein are set whatever gods there be,
like the branches of a tree about a trunk'. In the Mahabharata,
this imagery is changed to a lotus emerging from the navel of
Vishnu, holding Brahma, the Creator.

The word yaksha occurs several times in the *Rig Veda*,
Atharva Veda, Upanishads and Brahmanas. It generally means
'a magical being', something both terrible and wonderful.
The Mahabharata (II.5.100) enumerates the sacred trees of
various towns and tells us that every village had a sacred tree,
each identified with a yaksha. Each town had its own spirit
protector living in his particular tree, the vriksha chaitya.
Large towns like Rajagriha, Mathura and Benares had more
than one. Its frequent appearance in Sanskrit literature refutes
the many attempts to brand yakshas as non-Aryan. They are a
lower stratum of deities who were propitiated to prevent them
from doing mischief. Several Vedic gods, including Varuna
and Brahma, are also referred to as yakshas.

The worship of the yaksha in the tree was as important as
the worship of the tree itself. Trees were the natural abodes
of spirits, many of whom were identified by or named after
the tree. There is an entire Buddhist Jataka—*Rukkhadamma*
(or *Vriksha Dharma*) *Jataka*—dedicated to the worship of trees,
shrubs, bushes and plants.

A very important feature of yakshas and yakshis was their association with fertility. The Mahabharata refers to goddesses born in trees, to be worshipped by those desiring children. According to the epic, Vishvamitra and Jamadagni were produced by their mothers by embracing trees, while Queen Vitasavati of Bana's *Kadambari* worships the pipal and other trees in her desire for a child. Yakshas bestowed wealth and progeny; they were supposed to bring about marriages, guard women's chastity (*Rajatarangini*, I.319 ff), grant children and grandchildren (*Avasyaka Sutra*, II, p. 36) and protect the foetus (*Mahavamsa*, IX.22 ff). The *Hatthipala Jataka* (509) refers to a poor woman who points to the banyan tree as the source of her seven sons. Even today, the sacred tree of a temple or village is circumambulated by those who want progeny.[13]

The yakshas are vegetation spirits with powers of life. On a Bharhut coping, a vriksha devata—or yaksha—is seen as offering a bowl (of food) and a *kamandalu* or kettle from which water is poured into the hands of a man seated below. This is probably an illustration of the 'Story of the Treasurer' in the *Dhammapada Asthakatha* (I.204). This illustration reappears on a railing pillar in Bodhgaya. A man stands in front of a tree, receiving a kettle and a plate from two hands emerging from the tree, which has an altar beneath it. In a detail from the *parinirvana* from Mathura, a vriksha devata is seen standing hidden in a tree.[14]

In Jainism, yakshis look after the well-being of the Tirthankaras. Each Tirthankara has two attendants, a yaksha and a yakshi, with supernatural powers. Sometimes, they are the objects of worship too. The important Jain yakshis include Ambika (Amba, Amra), named after the mango tree, who was an attendant of Neminatha; Chakreshwari, attendant of Adinatha (Rishabhanatha); Kushmandi, who holds a mango; Lakshmi, holding a lotus, representing wealth; Padmavati, holding a fruit, attendant of Parshvanath; and Sarasvati, seated on a lotus.

Yakshas and yakshis are also protective guardians who flank the gateway in Sanchi. In later art, they were identified with the river goddesses Ganga and Yamuna who flanked the temple doorways. Finally, they became the goddesses Lakshmi and Sarasvati, seated on the sacred white lotus, a divine symbol of purity. Yakshas and yakshis stand under the *ashoka*, champaka and kadamba trees in Mathura art.

The *Bhagavad Gita* says that only persons of rajas quality worship the yakshas while the Buddhist *Maitrayini Samhita* (VII.8) and *Lalitavistara* attack the worshippers of yakshas. Jain texts also attack yakshas and yaksha worship. However, several Sanskrit texts, including the *Kathasaritsagara*, affirm that all sections of society—Brahmins, Kshatriyas, Vaishyas, Sudras and tribals—worshipped them.

The worship of the yaksha occasionally involved offerings of blood, flesh and wine, as described in the *Dummedha Jataka* (50), among others. Even human sacrifices were made to yakshas, according to the *Mahavamsa* (V.7.2) and the *Mahasutasoma Jataka* (V.258). The former describes the conversion of the cannibal Yakkha Ratakkhi, to whom *bali* (offerings of flesh) were made.[15]

However, with the gradual triumph of ahimsa, the worship of the individual tree, devoid of the yaksha, becomes more vegetarian and sattvic, with pradakshina (circumambulation), fruit and flowers. While the Buddhists sought to change the worship of the yaksha to a more sattvic form of worship and the Jains rejected it outright, Brahmanic religion retained the sacred tree alone and discarded the bloody sacrifices.

On the reliefs of the Bharhut, Sanchi and Amaravati stupas, there are trees associated with female figures—the yakshi or spirit of the tree. The *vrikshaka* (tree maiden) or *shalabhanjika* (a woman who breaks the *shala* [sal] tree) touches the tree—ashoka, *amra*, kadamba or champaka—with her foot for it to flower. She is responsible for the flowering and fruition of the tree whose branch she holds. The concept of the shalabhanjika is first mentioned in the 'Vanaparva' of the Mahabharata, where she is frequently addressed as *pushpa-shakha-dhara* (one who holds a flowering branch), and later in Asvaghosha's *Buddhacarita* (first century CE) and the 'Shrishtikhanda' (17.57) of the *Shivamahapurana*. The vrikshaka or shalabhanjika is a yakshi who resides in the tree. The fertility so gained by the tree became, in time, an attribute and ability of the tree itself. The *ashoka dohada* is a favourite practice in which the woman kicks the ashoka tree and makes it bloom. Kalidasa describes this ritual in *Malavikagnimitra* and *Raghuvamsa*. The tree is still popular in southern and eastern India and is even now associated with vrikshikas, also called *surasundari*s. The most common trees are the mango and the ashoka. The dryad motifs are closely connected with fertility. The best known and most spectacular yakshis are the shalabhanjikas of the Sanchi gateways—nude, opulent, sensuous figures holding the kadamba, ashoka or mango trees in full bloom, laden with fruits and a thick canopy—very obvious figures of fertility, a dichotomy in an otherwise monastic and near-puritanical situation.

Was it the woman or the spirit who was associated with the tree? A sculpture from Sanchi suggests the latter. A yaksha stands holding ripe mangoes on one side, with kadamba leaves hanging over his other shoulder. Obviously, it was the spirit of the tree that was immortalized in the sculpture. In some parts of India, there is still an annual ritual where women embrace the tree and partake of its bark or flowers.

The inscribed figures of the yakshas and yakshis, nagarajas and devatas found on the *torana*s and Jatakas constitute an extensive iconography of the tree and its associations.

The association of the tree and the snake appears on Harappan seals: two snakes flanking a tree, a naga-like figure guarding the shami, and so on. Much later, in the Bharhut images of the Sunga period

Figure 4.6: Snake stones, Kachapeshwarar Temple, Kanchipuram

Photograph by R. Selvapandian

(second century BCE), the nagas—men and women with cobra hoods —worship symbols of the Buddha placed beneath the bodhi tree. It is a common sight in rural India to see snake stones installed in front of trees (Figure 4.6), particularly the pipal, which is the most sacred tree of India. It is found in almost every village in the country and is believed to be the dwelling place of the Hindu trinity: Brahma, Vishnu and Shiva.

Kalpavriksha

The kalpavriksha, also known as *kalpataru*, *kalpadruma* and *kalpapadapa*, is the divine tree of life, guarded by flying *kinnaras*/*kinnaris*, apsaras and gandharvas. It is a popular motif in art and has even been adopted as an Islamic art motif.

The kalpavriksha is the wish-fulfilling tree, first mentioned in the *Rig Veda* (I.75; XVII.26). It came out during the churning of the ocean, and Indra, the king of the heavens, took it away to paradise. It is a very special tree that can bear any kind of fruit. It is a symbol of life and prosperity, a focal point for one's spiritual quest.

According to the 'Bhishmaparva' (7.2) of the Mahabharata, the Siddhas who lived in Uttarakuru worshipped the tree. Some branches produced streams of milk tasting like nectar and the flavour of all six rasas. Other branches produced clothes and ornaments, still others yielded men and women of great beauty and youth. It was a golden tree of life, with golden branches and fruit, and had to be worshipped properly.

A 200 BCE pillar from Vidisha, Madhya Pradesh, is carved like a banyan tree, and a conch, a lotus, a vase of coins and a bag tied with a string are hung from it. This was obviously the kalpavriksha of ancient legend. While the bodhi tree symbolizes enlightenment, the kalpavriksha is surrounded by a railing with gold coins hanging from the branches, symbolizing Lakshmi, the goddess of prosperity.[16] Several trees are known as the kalpavriksha—banyan, parijata, pipal and coconut tree (called kalpataru in the coastal districts). It is a sacred tree that bestows prosperity.

The kalpavriksha focuses the mind in a spiritual direction. It gives and never disappoints, whether they are spiritual or materialistic demands. However, it never grants evil desires. It is a means to reach the gods. Unlike human beings who ask for and desire material objects, the kalpavriksha, like all trees, gives food, shade, fruits, nut and timber, besides cleaning and purifying the air. It protects the traveller against heavy winds and stormy rains and is self-sufficient. It is an example for human beings: just as a tree bows before a storm out of humility and survives, we too must submit to Divine Will. If we perform good deeds for the tree by watering it and taking care of it, it will reciprocate by providing fruit and flowers and a safe haven. Travellers are encouraged to sleep beneath a kalpavriksha, for it is an excellent source of clean air, herbs and good aroma.

Chaitya Vriksha and Sthala Vriksha

The term *chaitya* denotes a tree with dense foliage and fruits which provides shelter. The tree was the chaitya or protector for sacred images before temples were built, hence the name. The yakshas were believed to live in the sacred trees, from where they could observe and participate in daily activities. In the fifth-century BCE Gupta temple at Deogarh, Nara and Narayana are depicted seated beneath the sacred badari tree. In the seventh-century BCE Jain temple at Aihole, the deity is standing beneath a *Cassia fistula* tree, flanked by a woman attendant on either side while two yakshas are seated on the tree. The Ramayana (II.3.18, 50.8; V.12.18, 22–29) says chaitya vrikshas were so called because the trees were planted with revetments around their stems. In the course of time, all such trees were called chaitya vrikshas. Almost all villages had their own chaitya vriksha. Such holy trees could not be harmed because they were the abodes of devas, yakshas and rakshasas (Mahabharata, I.138, 25).

When and why did the chaitya vriksha become the sthala vriksha? As the characteristics of the yaksha protector were absorbed by the tree, it became the protector of the town, village or *sthala*. Earlier, the tree was the chaitya or protector of the image of the deity. The emphasis in Hinduism, Buddhism and Jainism was on the images housed either in a rock-cut cave or beneath a tree. From the Puranic and historic periods onwards, the emphasis was on the construction of impressive temples. The protector (chaitya) tree gave way to the tree of the sacred place (sthala). Puranic literature is replete with mentions of the importance of pilgrimages to sacred sthalas. When a devout Hindu visits a temple, he will invariably visit and worship at the sthala vriksha, make offerings of flowers and coconuts to the tree or the snake stones beneath it. It is an integral part of the temple complex and rituals.

While sthala vrikshas abound all over the country, the concept of the chaitya vriksha has disappeared totally. Sthala vrikshas are associated either with the temple or the deity. Thus Shiva is generally associated with the bilva and Devi with the neem or kadamba. However, there are often aberrations to the rule and Shiva is associated with the Alexandrian laurel and mango at Mylapore (Chennai) and Ekamreshwarar (Kanchipuram), respectively. Each sthala purana has a legend linking the vriksha to the local deity or consort. Thus it was beneath the Alexandrian laurel and mango trees that Devi performed penance to regain the hand and heart of Shiva, in Mylapore and Kanchipuram, respectively.

The sthala vriksha may stand either on the ground or, more often, on a raised platform. It is generally situated in the outer prakara of the temple and is not part of the daily ritual. It is worshipped with offerings of coconuts, flowers and lighted lamps. People tie strings to the tree to fulfil a vow or a wish, generally to be blessed with children. The sthala vriksha is circumambulated by the worshipper who prostrates before the tree and sometimes encircles it with a string, as if to trap the spirit within.

Shiva as the teacher Dakshinamurti sits beneath the banyan, facing south, to enable his students to face north, the source of wisdom. This is a remnant of the earlier traditions of locating the deity beneath the tree.

While the sacred tree was called chaitya vriksha till the epic period, the term sthala vriksha appears in Puranic literature.

The *Matsya Purana* and *Padma Purana* describe various sthala vrikshas and *vriksha mahotsavas* or tree-planting festivals. It is interesting to observe the gradual disappearance of one term and the appearance of the other, as forms of worship changed. The sthala vriksha is associated with the sacred *sthalam*, *sthanam* or holy centre of pilgrimage.

According to the *Skanda Purana*, the devas worshipped the swayambhu lingas in the forest. The tree under which the linga appeared was described as the sthala vriksha. This legend points to the fact that the trees protected the image to be worshipped prior to the construction of temples. The oldest Hindu sculpture of the third century BCE, the Shiva linga at Gudimallam, once stood beneath a tree, as the stone *harnika* (fence) suggests.

Trees that have given their names to towns include the vrinda or sacred basil of Vrindavan; badari or Indian plum of Badrinath; *nel* or rice of Tirunelveli; and *mullai* or jasmine of Tirumullaivayil, among others. An ancient pipal tree at Joshimath in Uttarakhand is associated with Adi Shankara, who wrote the *Kalpa Shakti Sthavam* in praise of this divine tree. In Punjab, several gurudwaras are associated with trees, such as the *ber* or Indian plum (*Zyzyphus mauritiana*), beneath which Guru Nanak attained enlightenment. Guru Nanak described the tree as a saviour of creation and said, 'O God, you are an infinite tree and I am a bird under thine protection.'[17]

The worship of the spirit or yaksha is not forgotten. He continues to wander around the outskirts of towns and villages, a lost soul who mounts the terracotta animals offered to him, in the sacred groves where people believe he still lives.

Thus the sthala vriksha of contemporary Hinduism is a continuation of the simple tree-and-spirit worship of ancient India. It represents the sanctity imbued in ancient associations, a reminder of the role played by nature. It is a remnant of ancient pantheism, when people recognized the role of plants and their importance. It is a memory of the ecological heritage of the Indian subcontinent, where divinity was recognized, respected and revered in all life forms.

Plants, especially the sacred trees which are ecologically important, also provide 'ecosystem services'. They constitute a part of the genetic resources essential for the conservation of genetic diversity. They provide food, medicine, timber, non-timber forest products and water. They regulate the climate, control carbon sequestration, maintain the hydrological balance, rainfall and temperature and prevent diseases. They support nutrient cycles, pollination and biodiversity. They provide spiritual, cultural, aesthetic and recreational benefits. The ecosystem services of trees and forests are very important for society and community, but no serious study has been conducted on them. They also provide significant economic services to society in many different ways.[18]

Indian scientists have been working with Hindu priests at the major pilgrimage shrine of Badrinath in the Himalayas to encourage pilgrims to plant seedlings for reasons connected to their religious and cultural traditions. They hold planting ceremonies that allow people to enrich their pilgrimage experience by restoring an ancient sacred forest. Temples like that of Venkateshwara at Tirupati and Meenakshi at Madurai have a scheme of *vriksha prasadam*, whereby a devotee can pay for the planting and maintenance of one or more tree saplings for a divine blessing. This has proved far more efficacious than mere advisories to plant trees. Today Venkateshwara Vana is a thick green forest, where many kinds of wildlife, including predators, have made their home.

Only trees can save the planet, halt global warming and climate change, absorb the carbon dioxide spewed by a growing population of people and cattle, ensure the respiration that will give rain, and provide food and shelter to humans and so many mammals, birds and reptiles. If we can rekindle the faith people had in the sanctity of trees, maybe our planet will become a healthier place.

5

Children of Pashupati[1]

No person should kill animals who are helpful to all;
By serving them one should obtain heaven.

—*Yajur Veda,* XII.47

India's greatest contribution to world thought is the concept of ahimsa or non-violence, in thought, words and action. Killing animals has been prohibited since the Vedic period.

The Vedas and Upanishads were the first to speak of ahimsa. Although the Aryans were not vegetarians, the concept of non-killing appears in the earliest literature. The term 'ahimsa' is an important spiritual doctrine shared by Hinduism, Buddhism and Jainism, which implies the total avoidance of harm to any living creatures by thought, word or deed. Ahimsa has been described as a 'multidimensional concept', inspired by the belief that the Supreme Being lives in all living beings—human or animal. Therefore, to hurt another being is to open oneself to possible karmic repercussions. Ahimsa reached an extraordinary status in the philosophy of Jainism, which is based on the Upanishadic principle of *ahimsa paramo dharmah*. More recently, Mahatma Gandhi used ahimsa as a weapon to fight the trigger-happy British.

The *Rig Veda* (X.87.16) condemns all forms of killing, even for food, preferring vegans to drinkers of milk:

The *yatudhana* who fills himself with the flesh of man,
He who fills himself with the flesh of horses or of other animals,
And he who steals the milk of the cow:
Lord, cut off their heads with your flame.

The gods were called 'bulls', for the animal possessed the twin characteristics of manliness and gentleness.

According to the *Atharva Veda* (XII.1.15), the earth was created for the enjoyment of bipeds and quadrupeds, birds, animals and all other creatures, not humans alone. The *Mundaka Upanishad* (2.1.7) describes the emergence of all life forms from the Supreme Being:

'From Him, too, gods are produced many fold, the celestials, men, animals, birds . . .'

Rig Veda (X.22.25) uses the words *satya* (truth) and ahimsa (non-violence) in a prayer to Indra. The *Yajur Veda* says, 'May all beings look at me with a friendly eye, may I do likewise, and may we look at each other with the eyes of a friend.'[2] Ultimately, ahimsa becomes the concept that describes the highest virtue in the late Vedic era (about 500 BCE).

The term also appears several times in the *Shatapatha Brahmana* in the sense of 'non-injury'. The earliest reference to the idea of non-violence to animals (*pashu ahimsa*) is in the 'Kapisthala Katha Samhita' (31.11) of the *Yajur Veda*. Ahimsa as an ethical concept started evolving in the Vedas and became increasingly central in the Upanishads.

The *Chandogya Upanishad* (8.15.1) bars violence against 'all creatures' (*sarvabhuta*) and the practitioner of ahimsa is said to escape from the cycle of birth, death and rebirth, the transmigration of the soul of a human being or animal into a new body of the

same or a different species. It also names ahimsa, along with
satyavachanam (truthfulness), *arjavam* (sincerity), *danam* (charity)
and *tapah* (meditation), as one of the five essential virtues (3.17.4).
The *Shandilya Upanishad* lists ten virtues, of which ahimsa is one.[3]

The Mahabharata (XIII.117.37–38) emphasizes the
importance of ahimsa in Hinduism: Ahimsa is the highest
virtue, the highest self-control, the greatest gift, the best
suffering, the highest sacrifice, the finest strength, the greatest
friend, the greatest happiness, the highest truth and the greatest
teaching. It is discussed in several parts of the epic, particularly
in the famous dialogue between Yudhishthira and the dying
Bhishma. The epic says that 'all creatures act according to
the laws of their specific species as laid down by the Creator.
Therefore, none should act unrighteously, thinking, it is I who
is powerful' ('Vana Parva', 25.16).

The principle of the sanctity of life is clearly ingrained in
the Hindu religion. Only god has absolute sovereignty over all
creatures, including man; man has no dominion over his own
life or non-human life. Mankind cannot act as a viceroy of
god over the planet, nor assign the worth of other species. No
damage may be inflicted on another species. All lives, human
and non-human, are of equal value and all have the same right
to existence. The *Atharva Veda* (XII.1.15) says:

> Born of Thee, on Thee move mortal creatures;
> Thou bearest them—the biped and the quadruped;
> Thine, O Earth, are the five races of men, for whom,
> Surya (Sun), as he rises spreads with his rays.

Hinduism is noted for its respect and consideration for the
natural world. This includes the flora and fauna of the earth, and
creatures in the sky and under the sea. All that exists has been
created by the Supreme Being, comes from the Supreme Being

and will return to the Supreme Being, which is the basis for the veneration of the natural world in which man finds himself.

The *Manu Samhita* (5.45) says, 'He who injures innocent beings with a desire to give himself pleasure never finds happiness, neither in life nor in death.' The *Srimad Bhagavata Mahapurana* (1.7.38) says that a cruel person who kills others for his existence deserves to be killed, and cannot be happy, either in life or in death. The consequences, according to the *Yajnavalkya Smriti* ('Acharyadhyaya', 5.180), are that 'the wicked person who kills animals which are protected has to live in hellfire for the days equal to the number of hairs on the body of that animal'.

In the Puranas, killing animals and eating meat were considered to be such heinous crimes that neither prayers nor pilgrimages or bathing in holy rivers would absolve the consumer.

Although Hinduism did not require its adherents to be vegetarians, vegetarianism was recognized as a higher form of living, a belief that continues in contemporary Hinduism where it is considered essential for spiritualism and the liberation of the soul (moksha).

The arguments in favour of non-violence to animals refer to the happiness one feels, the rewards it brings before or after death, the dangers and harm it prevents and the karmic consequences of violence.

Around the sixth century BCE, two great religious preachers took the Upanishadic philosophy of good conduct and non-killing to the people in the common language: Mahavira the Jina (victor) and Gautama the Buddha (wise). Both emphasized the importance of ahimsa or non-violence. Their influence was so powerful that even the Vedic religion gave up sacrifice and vegetarianism became synonymous with the upper castes.

The ancient Hindu texts discourage the destruction of nature, including of wild and cultivated plants. Rishis and sanyasis were expected to live on fruits and nuts in order to avoid the destruction of plants. Thus the principles of ecological non-violence is innate in the Hindu tradition, and its conceptual fountain has been ahimsa as the cardinal virtue.[4]

The concept of ahimsa exists in the classical Tamil language too. The *Tirukkural,* written about 200 BCE and sometimes called the Tamil Veda, dedicates several chapters to the virtues of compassion and ahimsa, particularly vegetarianism, the non-harming of animals and non-killing of all life forms.

> What is compassion, and the lack of it: not killing and killing; it is not virtuous to eat meat obtained by killing (26.254).

> The jewel of the eye is compassion; without it eyes are but sores.
> The world is his who does his job with compassion (58.575–78).

> How he be kindly who fattens himself on others' fat?
> Better than a thousand burnt offerings is one life unkilled, uneaten.
> All living things will fold their hands and bow to one who refuses meat (26.251, 259, 260).

There are many more such couplets dotting the *Tirukkural* on compassion and vegetarianism.

Albert Schweitzer said that 'there hardly exists in the literature of the world a collection of maxims in which we find so much of lofty wisdom. Like the Buddha and the *Bhagavad Gita*, the [*Tiruk*] *Kural* desires inner freedom from the world and a mind free from hatred. Like them it stands for the commandment not to kill and not to damage. There appears in the *Kural* the living ethic of love.'[5]

In the Hindu tradition, animals are recognized as having feelings and passions as human beings. They can also understand human speech, thus becoming divine. By recognizing the divinity in animals, Indian religions gave them a unique position which helped protect many species. The deification of several animals led to their protection, a safeguard that was lost in the medieval, colonial and post-colonial period when many animals were described as vermin and hunted to death. The cheetah, derived from *chitra*, became extinct, while the lion, tiger and leopard came to near-extinction.

Animals were revered for several reasons. The elephant, a keystone species, was the remover of obstacles. This role of the animal in the Indian jungle was transferred to Ganesha. Hanuman, the langur, was a fellow primate. The fish or Matsya was an ecological indicator, and an incarnation of Vishnu. Varaha the boar was an indicator of rain and ploughed the soil, teaching and aiding the farmer. The cow was essential for milk. The bull was a draught animal. The blackbuck was essential for the survival of the khejri plant which was the mainstay of the desert. And many others, all of whom had an important ecological or social role. The elephant-headed Ganesha, Hanuman the langur, the animal incarnations of Vishnu, Vaghdeo the tiger and the blackbuck, among other animals, were deified for the qualities they possessed, and an elaborate mythology was built around each.[6]

> *Vidya vinaya sampanne brahmane gavi hastini*
> *Shuni chaiva svapakecha panditah samadarshinah*
> (*Bhagavad Gita*, 5.19)

Those who are wise and humble treat equally, the brahmin, cow, the elephant, dog and dog-eater.

Vehicles of the Gods

Many animals are considered the vehicles or vahanas or the
companions of the gods, sometimes even gods themselves. The
various cultural connections are expressed through myths and
religious practices that celebrate nature and natural resources.
The worship of each sacred element in nature reveals people's
knowledge of the connection between nature and spiritualism,
using religion to protect nature. Thus many animals became
vahanas or vehicles of gods. The bull and the eagle were
companions and vehicles of Shiva and Vishnu, respectively.
Many were totemic figures who were absorbed into the
Hindu pantheon. The totemic tradition was widespread in
ancient India: many Sanskrit *gotra* (lineage) names and names
of sages are of animal origin, such as Bharadwaja (owl), Garga
(crocodile), Rishyashringa (born of a doe), Jambuka (jackal)
and Gautama (rabbit). Many clan names have animal origins,
such as Maurya, More (peacock) and Ghorpade (monitor
lizard). The different belief systems coalesced into Hinduism.

Some animals were friends and companions. The dog was
a wanderer and companion, characteristics that were replicated
in Bhairava's canine companion. Some were local chieftains,
like Sugriva the monkey king, who built the bridge to Lanka
and provided the army for Rama; Jambavan the wise bear,
who advised Rama and his army; and Jatayu the vulture, who
gave up his life trying to save Sita.

Some vahanas have specific names, others do not. Often a
god is associated with different vahanas at different periods and
in different situations. Indra's original companion was Sarama
the dog; later the white elephant Airavata became his vehicle
and sometimes he rides the horse Uchchaishravas. Thus the
vehicle is a symbol.

The gods and their vehicles are:

- Agni (Fire)—Ram
- Ayyappa—Tiger
- Bhairava (Shiva)—Dog
- Brahma (Creator)—Swan
- Brihaspati (Creator)—
  Elephant
- Buddha—Horse
- Chamunda (Devi)—Owl
- Chandi (Devi)—Lizard
- Chandra (Moon)—Ten
  horses, antelope
- Durga—Lion, tiger
- Ganesha—Mouse, rat
- Ganga (River)—
  Crocodile, fish

Figure 5.1: Bahuchara Mata on rooster, Ravi Varma Press

Source: The C.P. Ramaswami Aiyar Foundation

- Indra—Elephant
  (Airavata), dog (Sarama), horse (Uchchaishravas)
- Jagaddhatri—Tiger
- Kama (Love)—Parrot, crocodile
- Kamakhya (also Bahuchara Mata [Figure 5.1], a form of
  Devi)—Rooster
- Kartikeya/Murugan—Peacock (Paravani), rooster
- Ketu—Eagle
- Kubera (Wealth)—Mongoose
- Lakshmi (Prosperity)—Elephant, owl, peacock
- Manasa—Elephant, snake
- Pushan—Goat
- Rahu—Lion
- Rati—Pigeon
- Sarasvati—Swan, peacock

- Shani (Saturn)—Crow, vulture
- Shashti—Cat
- Shitala—Donkey
- Shukra (Venus)— Crocodile
- Shiva (Destroyer)—Bull
- Soma—Antelope
- Surya (Sun)—Chariot pulled by seven horses
- Ushas (Dawn)—Chariot pulled by seven cows
- Varuna (Ocean)— Crocodile, tortoise, fish
- Vayu (Wind)—Antelope
- Vishnu (Protector)—Eagle (Garuda), snake (Adishesha/ Ananta) (Figure 5.2)
- Vishvakarma— Elephant
- Yama (Death)—Buffalo
- Yamuna (River)—Tortoise

Figure 5.2: *Vishnu on Garuda* by Raja Ravi Varma

Source: The C.P. Ramaswami Aiyar Foundation

Some animals were a part of social history. Mahisha, the buffalo vehicle of Yama, was the deity of ancient India, with many kingdoms named after him, such as Mysore (Mahisha-ur) and Mahishmati. The buffalo was worshipped by the indigenous pastoral tribes of India. The war between Mahisha and Goddess Durga replicates the conflict between the buffalo-worshipping pastoral tribes and agriculturists* who worshipped the Mother Goddess. When the latter won

---

* Popularly known as Dravidians, although the term is devoid of any ethnic or linguistic identity.

the war, the former became a
demon. But Mahisha lives on
as the buffalo god of the Todas,
Maria Gonds and as the deity
Mhasoba in Maharashtra.[7]

The interrelationship of terms
like 'wild' and 'tame' are complex,
since the two live in a continuum.[8]
The elephant, a wild animal,
is tamed and, in cruel irony,
Ganesha's attributes are the goad

Figure 5.3: Ganesha

and the noose that are used to control and train the elephant
(Figure 5.3). The buffalo, a tame herbivore, becomes the demon
Mahisha when confronted by the Mother Goddess.

There are three paths (*marga*) to the liberation (moksha) of
the soul: *jnana* or knowledge of the illusory nature of life (maya)
is the highest; karma or action follows; bhakti or total devotion
or surrender to one's personal god is the third. A human being
can consciously choose his path. Animals too can rise above the
limitations of their birth and need not be subject to the cycle of
life, death and rebirth. They too can attain liberation.

The Incarnations of Vishnu

Among the various incarnations of Lord Vishnu, the first
four—fish, tortoise, boar and man-lion—are animals.

Avatar means 'one who descends'. Whenever dharma
or righteousness is in danger, Vishnu incarnates himself to
save the world from evil. These incarnations perform all the
three roles of creator, preserver and destroyer, for he destroys
evil and re-establishes dharma. In the *Bhagavad Gita* (4.7–8),
Vishnu promises to incarnate himself:

yada yada hi dharmasya glanirbhavati bharata
abhyutthanamadharmasya tadatmanam srijamyaham.
paritranaya sadhunam vinashaya ca dushkritam
dharma samsthapanarthaya sambhavami yuge yuge.

(O descendent of Bharata, whenever there is a decline
in religious practices, and a predominance of irreligion, I
descend myself. For the deliverance of the pious, for the
annihilation of the miscreants, to re-establish the principles
of religion, I appear, millennium after millennium).

This stanza is one of the main tenets of the Hindu religion.
Krishna promises to intervene and uphold dharma. All creation
is placed on an equal footing. The most important and unique
aspect of Hindu theology is the association of every species
with reincarnation.

The Supreme Being actually gets himself incarnated in
the form of various species. *Srimad Bhagavata Mahapurana*
(1.3.5) says, 'This form is the source and indestructible
seed of multifarious incarnations within the universe, and
from the particle and portion of this form, different living
entities like demigods, animals, human beings and others are
created.'

While the popular belief is that there are ten incarnations,
of which nine have come and gone, some books like the
Bhagavata Purana claim there are twenty-two. There is a natural
evolutionary process in the avatars, for Vishnu incarnates as
several species, taking progressively developing forms. The
first four forms are, in order, the fish or Matsya, who lives
in water; Kurma the amphibious tortoise, who lives in the
sea and on land; Varaha the boar, the four-legged mammal;
and Narasimha the half-man half-lion. By incarnating himself,
Vishnu reiterates that all creation—animal and people—are

equal.[9] While the fish or Matsya was an ecological indicator of the waters, the tortoise or Kurma was an ecological indicator of the ocean. If Varaha the boar taught the farmer to plough the field, the lion aspect of Narasimha was a celebration of the predator.

When the ocean was churned (samudra manthana), out came nine gems: the divine elephant Airavata, the divine cow Kamadhenu, the divine horse Uchchaishravas and Lakshmi, the goddess of prosperity. This story may be an allegory for ancient trade: the mythical Airavata was believed to have been born of the River Iravati (Irrawaddy) in Myanmar and the horse was probably imported from the Arabian peninsula. Only the cow was indigenous.

Kashyapa was a divine progenitor or Prajapati. According to the Ramayana, he was the seventh and youngest son of Brahma. The Mahabharata says he was the only son of Marichi, one of the six mind-born sons (*manasa putra*) of Brahma. Kashyapa married the thirteen daughters of Daksha, who gave birth to gods, demons and all other creatures. Of the thirteen, Krodhavasa was the mother of Kamadhenu and all cows and elephants; Vinata of Aruna and Garuda the eagles; Kadru of the nagas or snakes; and Sarama of all canines. By making Kashyapa a divine progenitor, gods, demons, humans and animals became siblings. There is an echo of the Harappan horned deity, surrounded by animals, and the Vedic Pashupati, lord of the animals, in the literary descriptions of Kashyapa.

Animals interact with the human world through myths, which explain why:

- The squirrel got its stripes
- The cow is special

- The eagle devours the snake
- The huge elephant is so strong yet so gentle and so on

In the course of Indian history, certain individuals have contributed to the elevation of the status of animals. Rama, the hero of the Ramayana, came across several animals and indigenous tribes, especially the Vanaras or monkeys who built the *setu* or bridge to cross over to Lanka; the squirrel who carried sand and small stones on his body to help build it; Jambavan the wise bear; Jatayu the bird who fought Ravana and was killed, and his brother Sampati. Rama was identified as an avatar of Vishnu and became one of the most popular deities of Hinduism. As the cult of Rama developed, his associate Hanuman attained divine status (Figure 5.4). Killing monkeys is considered a heinous sin even today.

Figure 5.4:
Hanuman, the
langur-faced
friend of Rama

Drawing by S. Prema, The C.P. Ramaswami Aiyar Foundation

Several tribes—the Vrishnis, Ahirs and Yadavas—were followers of Krishna, who lifted the mountain Govardhana to protect the cowherds and their cattle. Krishna preached a religion of love for all creation. There was a special role in it for cows that gave milk—they were considered mother substitutes. This was to have far-reaching consequences.

Vegetarianism

Ashoka left inscriptions on rocks and pillars all over India urging people to have compassion towards animals and to ban the slaughter of cows and other lives.

The bhakti movement or belief in a personal deity started in Tamil Nadu in the early centuries of the present era, when the Shaivite saints spoke out against the caste system and animal sacrifice. Tamil Shaivism totally rejected meat-eating. Even today the Tamil word *saivam* means vegetarian. A sign of upward mobility in Tamil Nadu was a changeover from meat-eating to a vegetarian lifestyle.

Several medieval saints, like Ramananda, Mirabai, Kabir, Tulsidas, Surdas, Jnaneshwar, Namdev, Ekanath, Sant Tukaram, Ramdas, Purandaradasa, Kanakadasa, Vadiraja, Basavanna, Akka Mahadevi, Shri Krishna Chaitanya, Sankaradeva, Narasimha Mehta and Narayana Guru preached devotion to god, equality of man, kindness to animals and vegetarianism. Guru Nanak, founder of the Sikh religion, also preached vegetarianism. Swamis and gurus of contemporary Hinduism also spread the message of kindness to animals and vegetarianism, reinforcing in every age the idea that animals have souls and sentience.

The harsh environment of the Thar Desert in Rajasthan was the birthplace of Guru Jamboji, founder of the Bishnoi school, who prescribed eight principles to preserve biodiversity and encourage good animal husbandry. Jamboji died in 1536, leaving behind guiding principles for his community to save the environment and its animals. His principles became commandments: he preached kindness towards plants and animals, and said that he would be reborn in every blackbuck. Thanks to him, the Bishnois have never allowed anyone to kill any living being or cut any green trees. The success of their conservation efforts is seen in the green Thar Desert, covered with trees, where the blackbucks and chinkaras roam about freely and fearlessly.

About 40 per cent of Indians are estimated to be vegetarians. This would be the largest concentration of

vegetarians in the world. Indian vegetarians are mostly lacto-
vegetarians. Vegetarianism in India is dictated by religious
and caste traditions. Some, like traditional Bengali Brahmins,
include fish in an otherwise vegetarian diet, while others, like
the Jains, do not even touch root vegetables like garlic, onion
and potato.

Great Indian saints and seers like Vyasa, Panini, Patanjali
and Adi Shankara were vegetarians and spoke out against
eating meat, as they believed that right thinking and spiritual
attainment were not possible with meat-eating. According to
Hindu belief, eating meat and fish leads to negative karma.
Compassion towards the animals is essential for good karma.
Between Hinduism, Buddhism and Jainism, kindness to
animals and vegetarianism became hallmarks of an evolved
human being.

The first and third castes—Brahmins and Vaishyas—of
Hindus generally did not kill animals for food or sport. In Vedic
religion meat-eating was not banned but was restricted by
specific rules. Several scriptures bar violence against domestic
animals. The *Chandogya Upanishad* (8.15.1); Mahabharata
(III.199.11–12; XIII.115; XIII.116.26; XIII.148.17); and the
Bhagavata Purana (11.5.13–14) strongly condemn the slaughter
of animals and meat-eating.

The *Yajur Veda* adds, 'You must not use your God-given
body for killing God's creatures, whether they are human,
animal or whatever.'

According to the *Atharva Veda*, 'Those noble souls who
practise meditation and other yogic ways, who are ever careful
about other beings, who protect all animals are the ones who
are actually serious about spiritual practices.'

The *Manu Samhita* (5.48–52) recommends that since 'meat
can never be obtained without injury to living creatures, and

injury to sentient beings is detrimental to the attainment of heavenly bliss; let him therefore shun the use of meat. Having well considered the disgusting origin of flesh and the cruelty of fettering and slaying corporeal beings, let him entirely abstain from eating flesh.'

Further, it is not only the person who eats the meat but also the butcher and even the king or administrator who are at fault. The Mahabharata (XIII.115) contains a lengthy discussion about vegetarianism between the eldest Pandava prince Yudhishthira and his dying grandfather Bhishma:

> What need there be said of those innocent and healthy creatures endued with love of life, when they are sought to be slain by sinful wretches subsisting by slaughter? For this reason, O King, know that the discarding of meat is the highest refuge of religion, of heaven, and of happiness . . . The man who abstains from meat is never put in fear, O King, by any creature. All creatures seek his protection . . . If there were nobody who ate flesh there would then be nobody to kill living creatures. The man who kills living creatures kills them for the sake of the person who eats flesh. If flesh were regarded as inedible, there would then be no slaughter of living creatures. It is for the sake of the eater that the slaughter of living creatures goes on in the world. Since the life of persons who slaughter living creatures or cause them to be slaughtered is shortened, the person who wishes his own good should give up meat entirely . . . The purchaser of flesh performs *himsa* [violence] by his wealth. He who eats flesh does so by enjoying its taste; the killer does *himsa* by actually tying and killing the animal. Thus, there are three forms of killing. He who brings flesh or sends for it, he who cuts off the limbs of an animal, and he who purchases, sells, or cooks flesh and eats it—all of these are to be considered meat eaters.

Bhishma, grandfather of the Kauravas and Pandavas, continues, 'He who desires to augment his own flesh by eating the flesh of other creatures, lives in misery in whatever species he may take in his [next] birth.' (XII.115.47). Bhishma advises Yudhishthira: 'The man who, having eaten meat, gives it up afterwards wins merit by such a deed that is so great that a study of all the Vedas or a performance, O Bharata, of all the sacrifices [Vedic rituals], cannot give its like' (XII.115.16). In reply, Yudhishthira says, 'Alas, those cruel men who, not caring for various other sorts of food, want only flesh, are really like great rakshasas' (XII.116.1).

This conversation between Yudhishthira and Bhishma is significant in the preference for vegetarianism in Hinduism. The epic is clear that killing animals and selling and eating meat would earn negative karmas that would result in one's rebirth as a lowly being.

The *Bhagavata Purana* (7.15.7; 11.5.14) adds: 'Those sinful persons who are ignorant of actual religious principles, yet consider themselves to be completely pious, without compunction commit violence against innocent animals that are fully trusting in them.'

The Tamil *Tirukkural* (200 BCE–200 CE) has many references to vegetarianism, such as 'How can he practise true compassion who eats the flesh of an animal to fatten his own flesh?' Thus Thiruvalluvar implies that there was a vibrant discussion on vegetarianism and meat-eating in his times.

The Sacred Cow

The cow is so sacred in Hinduism that the term 'sacred cow' signifies an idea or institution held to be above criticism. The animal symbolizes dharma or righteousness. The reverence for

the cow has always been central to Hinduism. She is equated
to one's mother; because she gives milk she is the Gaumata
(cow mother). She represents both the mother and the earth.
She is the mother because her milk is the first replacement for
mother's milk and the earth because she is a symbol of fertility.
In times of distress, the earth is believed to take the form of a
cow to pray for divine aid.

The cow, descended from the zebu, is better known as
the humped cattle. She has lived in the subcontinent since the
beginning of life as we know it. She appears on the seals of the
Indus–Sarasvati civilization, among early terracotta figurines
and on rock paintings. In the Indus Valley seal of Pashupati,
the three-headed horned male figure surrounded by many
animals, the humped cow is present. Thus the sanctity of the
cow may be of very ancient origin.

Cattle were very important in the Vedic age. *Rig Veda*
(VI.28.1) says, '[The] cattle have come and brought good
fortune: let them rest in the cow pen and be happy near
us. Here let them stay prolific, many coloured, and yield
through many mornings their milk for Indra.' Her home is the
firmament or heaven (*Rig Veda*, III.55.1); she is Ushas or the
goddess of dawn; she is the earth, water and light.

The *Yajur Veda* (XIII.49) entreats the ruler not to kill
animals, and to punish those who do. 'O king. You should
never kill animals like bullocks, useful for agriculture, or like
cows, which give us milk, and all other helpful animals, and
must punish those who kill or do harm to such animals.'

The *Shatapatha Brahmana* (IX.3.3.15–17) says that '[the
cow is a] shower of wealth, the (cow's) body is the sky, the
udder the cloud, the teat is the lightning, the shower (of milk)
is the rain from the sky, and it comes to the cow'. The life-
giving rivers are compared to cows, for they are equally sacred.

Pashu means cow but also represents the world of animals, for Shiva as Pashupati is the lord of all animals, of which the cow is the foremost. Pashu is cognate with the Latin *pecu*, from which are derived words pertaining to money, such as *pecunia* (Latin) and impecunious (English). Cattle were essential to the Vedic economy. Wealth was estimated by the number of cattle owned by the individual. A cowherd (*gopalaka*) took the cattle out to graze and brought them back home in the evening, where they were housed in covered stalls. He was paid in kind, and was entitled to the milk of one cow out of every ten. The cows were milked twice a day except in spring, when the milk was reserved for the calves. Often quarrels broke out between villages over cattle.

There were strict punishments for killing a cow. The Vedic people were pastoral and depended on their cattle for their livelihood. Milk, fuel, fertilizer, medicines, disinfectants and anti-pollutants were all supplied by the humped cattle. The dowry and bride price were paid in cattle, while killing a cow was punishable by death. It was a heinous crime, on par with the killing of a Brahmin. The killer's head was shaved and he was fined ten cows and a bull (Mahabharata, III.240).

According to Puranic mythology, Brahma created the Brahmins and the cow at the same time; the Brahmins to recite the Vedas and the cow to provide ghee for the sacrifices. She is even addressed as the mother of the gods and Brahma declares that she should be worshipped.

The importance of the cow in Indian culture may be seen from the number of words derived from *gau* or *go*. One's ancestral family name is the gotra (or cow pen), within which the family lived with its cattle. So sacred is the gotra, or the male lineage, that two people from the same gotra cannot intermarry, even if one belongs to Kashmir and the other to Kanyakumari. The gateway to a temple is called *gopuram*, which means the village/town of the cow. Gauri, the consort of Shiva, is named after the Rigvedic buffalo-cow (*Rig Veda*, I.164.41).

During the shraddha ceremony (death rites), a cow must be gifted to a Brahmin, which will be received in heaven by the person who died. The cow will liberate the dead soul from all its sins.

The sanctity of the cow was so great that Babur, the first Mughal emperor, in his will to his son Humayun, advised him to respect the cow and avoid cow slaughter. The Mughal king Akbar chose to ban cow slaughter and thus endeared himself to his Hindu subjects. It is believed that the Indian Revolt of 1857 was prompted by the British colonizers asking the Hindu sepoys to bite a bullet that was greased with fat from cows (and pigs too, prompting Muslims to refuse, similarly, to bite the bullet).

Was the cow eaten during the Vedic period? This topic has become a subject of great controversy, with historians of the left and right taking opposing points of view depending on their political beliefs and not on what was actually written in the early Vedas. There is no doubt that some people killed cows and ate beef then as today. There is also archaeological evidence of knife marks on cattle bones. But does this mean that cows were killed for beef, or that killing the cow was sanctioned by any or all of the four Vedas?

The case for eating beef, or the flesh of the cow, rests on one word—*goghna*—which appears in the *Shatapatha Brahmana* and *Vashishtha Dharmasutra*, and which was translated by one Taranath, in the early twentieth century, as 'killer of the cow', from which it developed to 'serving beef to the guests'. However, this is at total variance with

Panini, the ancient grammarian, who translates the word as 'receiver of the cow' or one who receives a cow as a gift.[10] On the other hand, the *Rig*, *Sama*, *Yajur* and *Atharva Veda*s say that the cow is:

> *Aghnya*, one that ought not to be killed;
> *Ahi*, one that must not be slaughtered;
> *Aditi*, one that ought not to be cut into pieces.
>
> > (*Nighantu* [11.4] by Yaska, a commentator
> > on the Vedas)

The terms aghnya, ahi and aditi are also synonyms for gau.

The *Yajur Veda* is replete with instructions to protect the cow: 'Do not kill the cow which is the splendour of life and inviolable' (XIII.43) and 'You men and women, both of you together protect your cattle' (VI.11).

The *Atharva Veda* (VIII.3.25) adds: 'A man who nourishes himself on the flesh of man, horse or other animals or birds or who, having killed untorturable cows, debars them from their milk, O Agni, the King, award him the highest punishment or give him the sentence of death.'

There are many more references where killing the cow has been specifically banned. However, there is no doubt that many members of a primitive tribal society that existed five thousand years ago would have eaten beef. However, the question is whether Sanatana Dharma promoted it. The answer is definitely no.

By the epic period, all followers of Sanatana Dharma, had given up eating beef. There is enough literary evidence to prove that the Vedic religion did not permit the killing of cows or the consumption of its flesh. With a large number of texts giving a special status to the cow, it is impossible to believe that the Vedic people ate beef.

One of the defining features of medieval Hinduism was the ban on cow slaughter and the refusal to eat beef. It was a common criterion which rallied Hindus. In the fourteenth century, during the rule of Sikandar in Kashmir, masses of people were forcibly converted to Islam by being forced to eat beef. When they wanted to reconvert to Hinduism, they were not accepted by the local Brahmins. The failure of the Hindu clergy to relax the criteria that defined a Hindu—and refusing to accept eaters of beef back into the religion—forced many people who were converted to continue as Muslims. There are many examples of Hindus being forced to eat beef as part of the process of conversion to Islam by the invaders and Muslim rulers of medieval north India. They were excommunicated by their caste and could not return to their original religion.

The origin of vegetarianism in India has been linked to the veneration of the cow, which originated from the Vedic people whose sacred Vedas call for non-violence towards all bipeds and quadrupeds. Often, the killing of a cow is equated with the killing of a human being, especially a Brahmin. There is no doubt that the influence of the Upanishadic rishis contributed to the total ban on cow slaughter and beef-eating.

There are many special cows: Kamadhenu, the wish-fulfilling cow, born of the divine progenitor Kashyapa and his wife, Krodhavasa, a daughter of Daksha; Nandini (delight); and Surabhi (fragrant). Kamadhenu was the mother (Aditi) of all cattle, and one capable of granting any wish to the true seeker. It is believed that Surabhi worshipped Brahma for a thousand years, after which she was blessed with divinity. She became Kamadhenu the goddess and presided over Goloka, the heaven of cows. She is revered as a fountain of milk and one who grants every desire. All cows are described as

Surabhi's children, but she is also described as the child of Kamadhenu (Figure 5.5).

Surabhi is sometimes identified with Kamadhenu. In the Mahabharata, Kamadhenu comes out of the ocean during the churning of the ocean, when the devas (gods) and asuras (demons) churns the celestial ocean of milk for amrita, the nectar of immortality. However, according to a later story, Kamadhenu belonged to the sage Vashishtha. She was stolen by his student Satyavrata, who killed and ate a part of her, giving the remaining flesh to

Figure 5.5: Kamadhenu or the sacred cow

Source: The C.P. Ramaswami Aiyar Foundation

King Kaushika's family. Satyavrata was cursed by Vashishtha for three heinous sins: killing a Brahmin, stealing from one's teacher and killing a cow. Surabhi, the cow, belonged to sage Vashishtha and was coveted by King Kaushika. Vashishtha refused to part with her and the cow too refused to leave his side. This led to a terrible feud till King Kaushika realized the spiritual power of the sage, renounced his kingdom and became sage Vishvamitra.

Nandini means one who delights. She was another cow of plenty, the daughter of Surabhi, and belonged to sage Vashishtha.

Sabala was an extraordinary cow who belonged to sage Jamadagni, father of Parashurama, Lord Vishnu's sixth incarnation. Sabala was coveted by King Kartavirya. When the sage refused to part with her, he was killed by the king. When Parashurama saw his dead father and learnt the cause of his death, he avenged himself by wiping out twenty-one generations of the Kshatriyas, till he was halted in his terrible mission by Rama, the seventh incarnation of Vishnu.

Lord Krishna is usually depicted as a cowherd. The Yadavas and Ahirs, the tribes to which he belonged, were cowherds. He is known as Gopala, 'the protector of cows', and Govinda, 'the cow keeper'. Krishna lived in Gokul (the family of cows), and his heaven is Goloka. The famous hill of Vrindavan, where he lived, is Govardhana.

Krishna is usually portrayed holding a flute, standing beside a cow. He persuaded the cowherds (gopalas) and cowherdesses (gopis) of Vrindavan to stop their worship of Indra and to worship Govardhana instead, for the mountain provided pasture for their cattle. To protect them from Indra's anger, which manifested itself as a terrible storm, Krishna raised the mountain Govardhana to protect the cowherds, cowherdesses and the cows. This imagery appears over the centuries in Indian art, from the Govardhana Giridhari cave temple at Mamallapuram to the exquisite Rajasthani and Pahadi miniature paintings. The worship of the cow gained impetus with the growth of the Krishna cult, for Krishna was the divine protector of cows. He rejected the sacrificial worship of the Vedic Indra and regenerated the traditional beliefs of worshipping mountains and cattle.

It is believed that there is no salvation or liberation for one who kills the cow and that he will rot in hell. Eating beef is the most heinous sin for a Hindu, Buddhist and Jain. However, the cow is eaten in most Buddhist countries and by Indian Buddhists, in spite of the Buddha's specified reverence for the cow.

The cow is worshipped on the first day of Vaishakha when, it is believed, Brahma created the cow. The horns are painted yellow or saffron and the cow is bathed by milkmen. In Tamil Nadu, the day after Pongal (or Sankranti, 14 January) is Mattu (cow) Pongal, when the cow is washed, her horns painted and she is worshipped in gratitude and reverence. *Godana* or gifting of a cow is an act of great religious merit.

Panchagavya—the five products of the cow, including cow's milk, curd, ghee (clarified butter), urine and dung—is purificatory and medicinal. Every produce of the cow is used in Indian medicine and religious rites. Milk is believed to have sattvic qualities. Butter is churned from milk and clarified into ghee, which is used in Hindu rituals and for preparing the sacred food offered to the gods. Cow dung is used as a fertilizer, as a disinfectant and as fuel for the kitchen and the sacred fire. Modern science has confirmed that cow dung smoke is an excellent disinfectant and anti-pollutant. Cow's urine is used as a medicine for treating cancer in humans, among other diseases, and as an insecticide. The animal is a symbol of health and abundance, and its image is used everywhere—from temples to advertisements.

The cow holds a special place in the Indian Constitution which recommends the banning of cow slaughter. However, this has to be implemented by the states, some of which (like Kerala and Bengal) have not done so. Further, only milch cows are protected in many states. Once it stops giving milk, its killing is permitted. Cow slaughter is banned in Nepal, once a Hindu kingdom. In many cities, they are abandoned when they stop giving milk. This way, the owners avoid both the expense of feeding them as well as the sin of killing a cow. Recently, under the Prevention of Cruelty to Animals (Regulation of Livestock Markets) Rules, 2017, issued under

Section 38 of the Prevention of Cruelty to Animals Act, 1960, by the Ministry of Environment, Forests and Climate Change, the sale and purchase of cattle from animal markets for slaughter was banned. The new rules are not about controlling the eating habits of people, but about animal welfare. Livestock markets are meant only for legitimate agricultural purposes. Slaughter markets must buy directly from the farms.

Cows are maintained by several temples to enable the worshippers to earn merit by feeding them. Unfortunately, the reverence for the cow has now led to intensive milk farming. India as the world's largest producer and exporter of milk has worked to the detriment of the cow which is forcibly impregnated and kept constantly pregnant and injected with oxytocin to produce more milk. Her calves, especially the males, are taken away from her. The male calves are either abandoned or suffocated or just allowed to starve to death. There are more than 200 million cattle in India, more than any other country in the world.

Old cattle that can no longer give milk have to be cared for till their death, according to Hindu religion, in pinjrapoles or retirement homes, maintained by Hindu or Jain religious trusts, temples and individuals. Where there are no pinjrapoles, cows are abandoned and left to fend for themselves till they die.

The cow was also protected for producing male calves, which, as adult bullocks, did the work of tractor, thresher and transport vehicles for the farmer. The prohibition against beef in fact includes the flesh of bulls and bullocks also, but only the cow is regarded as sacred today.

The bull is powerful, hence Indra and Agni are often likened to this animal. The bull is no less sacred. In a famous Harappan seal, a male figure, probably Shiva as Pashupati, is surrounded by several animals, one of them a bull. The gods of the *Rig Veda* are called bulls, depicting virility and strength. Nandi is the vehicle of Shiva. He is the chief of the *gana*s and his images are invariably present in Shiva temples. In many temples, such as those at Kailasanatha in Kanchipuram or Brihadishvara at Thanjavur or the Bull Temple at Bengaluru, the image of Nandi the bull may be as large as or even larger than the primary icon of the Shiva linga.

The nilgai or blue bull actually belongs to the antelope family. Its resemblance to the cow has led to the belief that it was a sacred animal.

Other sacred hoofed animals include the blackbuck, the vahana of Vayu the wind, which is sacred to the Bishnois of Rajasthan. The blackbuck is essential for the ecology of the desert, for it feeds on the leaves of the khejri tree and promotes its growth.

If the bull is the sacred mount of Shiva, Vishnu's eagle emblem symbolized the sun and Durga's lion emblem fertility. Durga's mount, the lion, was also the mount of the Babylonian Ishtar and the Greek Artemis, and probably came to India with migrants or travellers.[11]

There are other predators too that are regarded as sacred. The tiger is a mount of Shakti. While in Maharashtra, Vaghdeo is a god of the sacred forests, in Karnataka it is Betaraya, another tiger deity. In North Canara, the sacred forests are known as Hulidevaruvana or 'forest of the tiger god'. Kerala's Ayyappa, the lord of Sabarimala, probably the largest pilgrimage centre in the world, rides a tiger. In Bengal, both Hindus and Muslims worship the tiger, which is the scourge of the Sundarbans, where the Muslim Bonbibi rides the tiger.

Among the agrarian groups of Goa is the practice of *manngem thapnee* (crocodile worship). The mugger or marsh

crocodile is worshipped on the new moon day of the month
of Paush. Terracotta figures of the reptile are constructed and
decorated with clam shells, flowers and vermillion, while oil
lamps are lit before the figurines. The practice is rooted in
the belief that appeasing the rain god Varuna by worshipping
his mount, the crocodile, will ensure
protection against the inundation of
paddy fields following heavy rains.[12]

Bhairava is a form of Shiva as
the wanderer. Just as many Bairagis
keep dogs as companions, so does
Bhairava. The word Bhairava means
'terrible' or 'frightful'. The dog is a
faithful friend, whatever the avatar
of his owner. Sometimes Bhairava
even rides on the dog (Figure 5.6).
Dattatreya, the incarnation of the
holy trinity of Brahma, Vishnu and
Shiva, is accompanied by four dogs,
who represent the four Vedas, and the
sacred cow.

Figure 5.6:
Bhairava with his
canine companion

Eagle, the vehicle of Vishnu, is not
the only sacred bird. The cuckoo is the vehicle of Kama, the
god of love. The crow is the vehicle of Shani or Saturn and also
represents the souls of departed ancestors. The parrot is held by
several goddesses: Kanchi Kamakshi, Madurai Meenakshi and
Andal the Vaishnava poetess-saint. The magnificent peacock is
the vehicle of the general of the gods, Karttikeya, the son of
Shiva and Parvati.

The cobra, the denizen of the underworld, is at the head
of a cult of his own. He is wound around Lord Shiva's chest;
he is tied around Lord Ganesha's waist; and he is Lord Vishnu's

couch. But snake stones are found beneath pipal trees and near anthills, where they are believed to live. The worship of the snake was so prevalent in ancient India that an entire tribe of people—the nagas—was named after the snake. They were even represented by an artificial snake on their turban, as may be seen in sculptures from Bharhut to Amaravati.

The protection of these animals was essential for the ecology. The predators were essential for the health of the forests and the cattle for milk production. The elephant's role was to push aside hurdles in the forest and create pathways: as Vighneshwara, he removes all obstacles. The snake kept the underworld and undergrowth clean (of rats and subterranean creatures).

By giving animals a sacred position, ancient Indians recognized the divinity in all creatures, who are subject to the same laws of karma, birth and rebirth. It was also another way to protect the ecology and economy, for which animals are irreplaceable, whether it was the snakes which controlled rats or cows which provided milk or the tigers and lions that ensured the health of the Indian forest.

The Prevention of Cruelty to Animals Act, 1960, was passed by the Indian Parliament as the first comprehensive legislation in the world protecting the rights of domesticated animals, while the Wildlife Protection Act, 1972, banned the hunting of wildlife and established national parks for their protection. In 2001, the government passed the Animal Birth Control (Dogs) Rules by which dogs could not be killed and had to be sterilized and vaccinated against rabies. Leather, animal fat (used, for example, in soaps) and other animal-based products are slaughter house by-products and animals are not killed specifically for that purpose. However, the leather industry has grown considerably, leading to clashes between *gau rakshak*s and butchers.

Nowadays the growing pharmaceutical industry is using animals—dogs, macaques, rats, rabbits, etc.—and this has become an issue for animal activists. In 1977, India banned the export of macaques for research. Our country has also banned the testing of cosmetics and its ingredients on animals in 2013 and the import of cosmetics tested on animals in 2014.

Yet recently, an Ayurvedic company obtained permission to test its products on animals. It is ironical that Ayurveda, which is a traditional Indian medicine system based on sacred plants, is now going to be tested on animals. Thus, for every step forward, there is one step backward.

6

Abode of the Gods[1]

Revered by all, stands the massive mountain,
The divine Himalaya, in the north of India.
Running his course from the eastern seas to the western ocean,
He[*] provides a measure for the earth's dimensions.

Made into a calf by several mountains,
With Meru disguised as a cowherd, Himalaya helped
To milk the earth like a cow, who at King Prithu's[†] instance
Yielded brilliant gems, herbs and spices.

—Kalidasa, *Kumarasambhavam* (1.1)[2]

M ountains are figures of awe, revered as mysterious places with the power to evoke an overwhelming sense of the sacred. Regarded as images of the world's axis, they convey multiple ideas of the centre of the universe, and as such are venerated by all cultures.

Mountains may be considered sacred in several ways:

---

[*] Himalaya is imbued with human capabilities by Kalidasa.
[†] Prithu was a mighty king from whom the earth obtained her name Prithvi; also an incarnation of Lord Vishnu.

- Some are designated as sacred because of traditions, myths and beliefs. For example, Mount Kailas (in Tibet) is considered the abode of Lord Shiva.
- A mountain may be associated with individual gods or saints, or may contain sacred sites such as temples and groves. For example, Mount Govardhana at Vrindavan is revered for its association with Lord Krishna.
- Certain mountains that may not be considered sacred in a traditional sense are revered as places for spiritual attainment. For example, Arunachala Hill in Tiruvannamalai.
- Mountains are a source of water, life, fertility and healing. Hindus look up to the Himalayas as the source of sacred rivers, such as the Ganga.

Mountains played a vital role in the conservation of local ecology and environment, which is why they were deemed sacred. Many beliefs are part of sacred mountain traditions, which are an important link between cultural identity and traditional patterns of land conservation. Sacred mountains are distinguished from other sacred sites as being exceptionally comprehensive ecosystems. Due to their topographic location and bio-cultural richness, they provide opportunities for climate change adaptation and act as a refuge for plants and animals during environmental change and from competing species. They provide opportunities for species to move up and down and to adapt to climate change, which can play a vital role in the survival of species.[3]

Sacred mountains and sacred sites within mountains have resulted in communities maintaining and preserving their natural resources in often-pristine conditions. Indigenous communities have long realized the value of the high diversity and natural resources within mountains, which nurture precious resources of nature. The sacred mountain protected due to cultural beliefs has resulted in precious water, timber, flora, fauna and other natural resources being maintained and preserved for future generations.

Sacred mountains have a special value that makes them worth protecting at all costs. Beliefs and attitudes held by people who revere them can function as powerful forces to preserve the integrity of natural environments, promote conservation, restore damaged environments and strengthen indigenous cultures. These mountains highlight values and ideals that profoundly influence how people view and treat each other and the world around them. In order to be sustainable over the long term, environmental policies and programmes need to take such values and ideals into account; otherwise, they will fail to enlist the local and popular support that they need to succeed.

In India, most sacred mountains are associated with Lord Shiva or Goddess Durga, generally in the form of a local deity or incarnation. In the south, Karttikeya is often the deity whose temples are situated on top of mountains, while in the north, it is Shiva.

The Himalayas

Himalaya, the 'Abode of Snow', has been personified as a many-peaked mountain, the god Himavat. Parvati Haimavati or Uma, the daughter of the mountain, became the consort of the great Lord Shiva himself. Many gods have their abode here, together with semi-divine gandharvas (celestial musicians) and demons or rakshasas. The Himalayas are the source of the sacred River Ganga, another of Himavat's daughters. Several great shrines are situated here, such as Badrinath, consecrated to

Vishnu, and Kedarnath, consecrated to Shiva. Mount Kailas stands high above the rest, for it is Lord Shiva himself. The Himalayan range forms a beautiful arc from Kashmir through Nepal to Arunachal Pradesh (Figure 6.1).

Figure 6.1: Himalayas as pictured by NASA Landsat 7 satellite

Once, Parvati playfully covered Lord Shiva's eyes as he sat in meditation on Mount Kailas. Instantly all light and life were extinguished in the universe until, out of compassion, the god opened his third eye, which blazed like a new sun. So intense was its blaze that it scorched the mountains and forests of Himavat to oblivion. Only when he saw that Parvati was contrite did Shiva relent and restore her father to his former estate.

Lord Shiva is the Great Ascetic, and innumerable yogis and rishis are believed to dwell in the remote caves of the Himalayas, performing austerities. To see the Himalayas is to have one's sins wiped away.

Kailas

The landscape of the corner of the great plateau of Tibet in which Kailas is situated is one of desolate beauty. In the high altitudes prevailing there—13,000 ft and more—virtually no trees grow and little vegetation clothes the rugged terrain. Due to the transparency of the rarefied air, however, colours reach the eye with unfiltered intensity: rich reds, browns, yellows, purples—and in fine weather both sky and mirroring water are a deep, noble blue. Climate, on the other hand, is unpredictable, at times violent, and always prone to extremes of heat and cold. It is said that while a man's arm, exposed to the heat of the sun, may be getting scorched, his feet, lying in shadow, may at the same time be suffering the ravages of frostbite. Not surprisingly, therefore, this has always been a scantily populated area. The sacred Mount Kailas stands out of this elemental landscape, a compelling and uncannily symmetrical peak. Sheer walls of horizontally stratified conglomerate rock form a monumental plinth thousands of feet high that is finally capped by a cone of pure ice. Such is the regularity of the mountain that it looks as though it might have been carved by human— or more accurately, superhuman—hands: those of the gods in fact. Kailas has been frequently compared to a great temple, cathedral, or stupa, one of those characteristically Buddhist monuments known in Tibet as *chorten*. The analogy almost invariably has religious connotations, for in some mysterious ways Kailas seems to have the power to touch the spiritual heart of man; in the past this has been as true for hard-headed explorers as it has been for the more impressionable pilgrims.[4]

The most sacred face of Mount Kailas is the one facing south, which has three horizontal lines above (Shiva's *vibhuti*?), a vertical line for the nose and two more horizontal lines where the eyes should be. This is the face over Manasarovar, and is regarded as Dakshinamurti (facing the south), while the more beautiful face looks north (Figure 6.2).

For Hindus, Kailas is the home of Lord Shiva, the destroyer of evil, and is symbolic of

Photograph by author

Figure 6.2: Kailas—north face

Shiva's symbol Om. According to the *Vishnu Purana*, the four faces of the mountain are made of gold, ruby, crystal glass and lapis lazuli. It is a pillar of the world and is located at the heart of six mountain ranges, symbolizing a lotus.[5]

Every year, thousands of pilgrims visit Kailas by foot, an ancient tradition. Circumambulating the 52-km path around the mountain on foot is a holy ritual that is believed to wipe out one's sins and ensure a place in heaven. It is done in a clockwise direction by Hindus and Buddhists and anticlockwise by Jains and Bons. The mountain remains unclimbed due to its sacredness (with rumours floating around that a European and another who tried to climb it never returned). In any case, the Chinese have banned people from climbing the mountain in deference to religious sensibilities. The Tibetan name for Kailas is Kangri Rinpoche (precious mountain). It is believed to be the Axis Mundi (cosmic axis, world axis, world pillar, centre of the world, world tree) in Eastern beliefs and philosophies or the connection between heaven and earth. Symbolically, Kailas is likened to the divine world, separate from the earth.

The starting point of the Kailas parikrama is the Yama Dwar, or the gate of the lord of death. Yama is the deity who brings mortal souls to their onward journey. Yama Dwar is actually a chorten or stupa which is encircled thrice by the faithful to avoid rebirth. It is believed that one must abandon his or her mortal self before passing through it. Important stops on the parikrama include Diraphuk at a height of 4760 m and Zulthulpuk and Dolma La at 5400 m. The melting snows of Kailas come down as the River Karnali at Diraphuk. While Hindus, Jains and Bons walk around the mountain, Tibetan Buddhists do a *sashtanga namaskar* for each step around the mountain.

Just as Vaikunth is the heaven of Vishnu, Kailas is the heaven of Shiva, who sits in perpetual meditation with his consort Parvati, the daughter of Himalaya (Figure 6.3). In the *Mahanirvana Tantra*, where Lord Shiva expounds the principles of tantra to his shakti or consort, the mountain is described in the following terms:

Source: The C.P. Ramaswami Aiyar Foundation

The enchanting summit of the Lord of Mountains, resplendent with all its various jewels, clad with many a tree and many a creeper, melodious with the song of many a bird, scented with the fragrance of all the season's flowers, most beautiful, fanned by soft, cool, and perfumed breezes, shadowed by the still shade of stately trees; where cool groves resound with the sweet-voiced songs of troops of Apsara [heavenly nymphs] and in the forest depths flocks of kolila [cockatoos] maddened with passion sing; where [Spring] Lord of the Seasons with his followers ever abide . . . ; peopled by [troops of] Siddha [holy men of semi-divine status] Charana [celestial singers, dancers, bards or panegyrists of the gods], Gandharva [celestial musicians] and Ganapatya [devotees of the god Ganesha!⁶

Figure 6.3: Shiva and Parvati in the Himalayas, a painting by artist K. Madhava Menon

The Ramayana and the Mahabharata frequently refer to Kailas as a comparison for anything of great height. It is said to be six leagues high, an assembly place for gods and demons and the site of a great jujube tree.

According to the Puranas, Kubera, the god of wealth, ruled from a fabulous city called Alaka, which was situated near Kailas, and the eight lesser peaks nearby were his treasure houses. In Kalidasa's Sanskrit epic, *Meghadutam* (Cloud Messenger), a lovelorn yaksha banished from Alakapuri recruits a passing cloud to carry a message to his estranged wife, who still resides in Kubera's city. The poet speaks of Kailas being used as a mirror by apsaras: '[Y]ou will come to be the guest of Kailasa which plays the mirror for celestial ladies and whose joints of table-lands (plateaus) were loosened by the arms of Ravana and which spreading over the sky with the heights of its peaks, white like lilies, appears like the laughter of the three-eyed lord (Shiva) accumulating in heaps day by day.'[7] It is likely that the eight lesser peaks include the six which are regarded as the six faces of Shanmugha. The remaining two are Parvati Parvat, behind Kailas, and Nandi Parvat, facing the mountain.

Buddhists associate Kailas with a tantric meditational deity called Demchog or Chakrasamvara (Supreme Bliss) and his consort Dorje Phamo or Vajravarahi. Demchog is associated with two other Tibetan mountains besides Kailas: Lapchi, near Nepal, and Tsari, 200 miles east of Lhasa. The Buddha was also believed to have inhabited the sacred mountain with a retinue of 500 bodhisattvas. In modern times, the most important Buddhist association is with the great guru-poet Milarepa, who lived in the late eleventh and early twelfth centuries CE and belonged to the Karma Kagyu school of Tibetan Buddhism. Legend holds that Milarepa was involved in a vital struggle for possession of Mount Kailas with Naro Bon-Chung, a priest of the Bon faith. In Jainism, Kailas is called Astapada, where Rishabhanatha, the first Tirthankara, attained moksha or liberation. And finally, for the Bons,

Mount Kailas, called Tise, was the Soul Mountain of Zhang Zhung.[8]

Was Mount Kailas the mythical Mount Meru, the cosmic mountain or Axis Mundi of Hindu, Buddhist and Jain cosmology? In Hindu cosmology, Meru is the sacred mountain with five peaks, regarded as the centre of the physical, metaphysical and spiritual universe. The mountain's height is reputed to be 84,000 *yojanas*.* It is difficult to give an accurate equivalent for a yojana, which is a figurative expression denoting sheer vastness of height. Early Hindu cosmological notions have Mount Meru standing at the centre of a complex multidimensional system embracing both material and spiritual dimensions. The various heavens and hells are disposed in order of hierarchy along a vertical axis running through the centre of Meru, while the earth, a material dimension, is disposed along a horizontal plane extending outwards from the body of the mountain, below the median level. In some versions, the seven continents are shown radiating outwards in succession, each separated from the next by a sea. A curtain wall of mountains forms the outer boundary, beyond which is the void. In other versions, the earth resembles a great lotus flower, with the continents arranged like petals around the great central pericarp of the mountain. The sun, moon and other heavenly bodies take their orbits around Meru, and day and night are believed to be caused by the interposition of the mountain between the earth and the luminaries in the heavens. The Pole Star stands directly above its summit. Meru is the home of the highest of the gods—a kind of Hindu Olympus. It is principally associated with Brahma, whose palace and throne are situated on the summit. Other important deities have their

---

* About 10,82,000 km.

abodes elsewhere on the mountain. And finally, Meru is the
source of all the life-giving waters of the world. These rivers
are the Sita, the Alakanada, the Chaksu and the Bhadra.

Buddhism came to Tibet and took root in the seventh
century CE. The concept of Mount Meru would have been
brought to Tibet with the Buddha's teachings. Kailas was
deeply venerated by the followers of Sanatana Dharma long
before the advent of Buddhism. Once Kailas had become a
Buddhist sacred mountain, the connotations of Meru would
have been transferred to it. If such a development did take
place, it would have been consolidated by the fact that since
time immemorial, and subsequently, the Hindus who had
been making pilgrimages to Manasarovar began increasingly
to venerate Kailas and endow it with, among other things,
the connotations of Meru as laid down in their own scriptural
traditions. Hinduism and Buddhism are not mutually opposed
and contradicting religions but, rather, part of a single tradition
and tend to accept and even adopt each other's beliefs and
practices with an open-mindedness that is hard for westerners
to comprehend.

Mountains provided the prototype for the classical
Hindu temple, and Meru was the ideal. Every Hindu temple
is built on a cosmic plan, symbolically pierced by the Axis
Mundi, which emanates from well above the building, enters
it through the cupola, proceeds downwards through the
successively enlarging pyramidal superstructure and finally
penetrates the *garbha griha* or sanctum sanctorum. This is
the transcendental reality of purusha, which begins from an
unknown point above and passes through all the storeys of
the temple until it reaches the heart of the temple. Many
temples have been designed as symbols of Mount Meru,
which is a part of the cosmic ocean: the sun, planets and

stars are believed to circle the mountain as a single unit. Mount Meru is clearly mythical, but was the epitome of a sacred mountain.

The best recreation of Meru is the Hindu temple of Angkor Wat in Cambodia, with its five spires representing the five peaks, the (original) seven walls symbolic of the seven continents and interspersed with

Figure 6.4: Aerial view of Angkor Wat, Cambodia

Source: Shyam tnj, commons. wikimedia.org.wikimedia.org

moats that represent the oceans. Meru sits on Jambudvipa, the earth's landmass. To the south of Jambudvipa is Bharatavarsha. Heaven is on top of Mount Meru, the home of the devas, gandharvas, kinnaras, apsaras and other celestial beings (Figure 6.4). Yet another example is the Javanese Buddhist temple of Borobudur, built as nine stacked platforms, six square and three circular, topped by a central dome.

Mount Meru, the mythical mountain of gold, is considered to be the dwelling place of the gods and the central point of the universe. Brahma lives on the top; Ganga falls on its peak from heaven before flowing down to the earth. The mountain was considered to be the centre of all physical and spiritual universes in Hinduism, Buddhism and Jainism.

Mount Meru appears in several episodes in Sanskrit literature. The mountain was the pole around which the serpent Vasuki was wound, when the ocean of milk was churned for the divine nectar or amrita. Vishnu took the incarnation of a tortoise to hold Mount Meru on his back.

Mena, the daughter of Mount Meru, was the wife of Himavan. Mena and Himavan had two daughters: Ganga, who became the river, and Uma, the wife of Rudra (Ramayana, I.35.15–16).

According to the Mahabharata, the Pandavas and Draupadi climbed Mount Meru to reach heaven. But Draupadi and four Pandavas fell down along the way for their sins and died. Only Yudhishthira and a faithful dog managed to climb it successfully and reached heaven.

A number of mountains in Asia, such as Mount Kailas in Tibet and Gunung Agung in Bali, Indonesia, provide the pattern for the mythical Mount Meru or Sumeru, which stands as a cosmic axis around which the universe is organized.

The most sacred mountains are in and around the Himalayas.

Jammu and Kashmir

AMARNATH is situated in the Ganderbal district near Sonmarg in Kashmir. It forms part of the Himalayas, 6 km south of Zojila and 117 km north-east of Srinagar. There is a cave, believed to be one of the most sacred places of pilgrimage in Hinduism, at the south face at an elevation of 3800 m, visited by Hindu pilgrims in an arduous climb during summer. Inside the cave is a Shiva linga made of ice, which waxes and wanes with the moon. The Amarnath Mountain is unclimbed due to its sacredness.

SHANKARACHARYA HILL is a part of the Zabarwan Mountain that occupies the central part of Kashmir. This hill was known as Jetha Larak in olden times and was later changed to Gopadari Hill. It is believed that the renowned philosopher

Adi Shankara visited the Kashmir Valley over a thousand years ago and resided here for some time. The hill and its ancient Shiva temple were named after him. There is also a theory that the temple was originally built by a king Sandiman in the third millennium BCE and renovated by King Gopaditya who gifted it to the Brahmins of Aryavarta in the second millennium BCE.

HARI PARBAT, also known as Sharika Peeth, is a hill overlooking Srinagar. There is a famous Shakti temple on the western slope. The hill is considered sacred by Kashmiri Pandits due to the temple of Goddess Jagadamba Sharika Bhagawati, who is depicted with eighteen arms. According to local legend, Hari Parbat was once a huge lake inhabited by the demon Jalobhava. When the inhabitants called the goddess for help, she took the form of a bird and dropped a pebble on the demon's head. The pebble grew larger until it crushed him. Hari Parbat is revered as that pebble, and is said to have become the home for all the gods of the Hindu pantheon. Another version of the myth says that two demons, Tsand and Mond, occupied the valley. Tsand hid in the water near the present location of Hari Parbat and Mond somewhere above the present Dal Gate, and both terrorized the people of the valley. The gods invoked Shakti who assumed the form of a *haer* (myna) and flew to Sumer, picked up a pebble in her beak and threw it on the demon Tsand to crush him. The pebble grew into a mountain and was therefore named Hari (myna) Parbat.[9]

Himachal Pradesh

MANIMAHESH KAILAS, also known as Chamba Kailas, is situated in the state of Himachal Pradesh. It towers over Lake

Manimahesh and forms the watershed of the Bhudil Valley which is a part of the mid-Himalayan range. It is the source of numerous rivers and streams: River Manimahesh Ganga, River Dhancho, and the streams Siv Korotar and Gauri, all of which join the Bhudhil River, which is venerated by the local people. In the month of Bhadon, on the eighth day of the new moon, a fair is held near the lake where, it is believed, Lord Shiva and Goddess Parvati come to take a bath. Lakhs of people from all over the country come here to bathe in the lake. Devotees believe that if bad weather covers the peak with clouds, it is a manifestation of divine displeasure. In one popular legend, Lord Shiva created Manimahesh after he married Goddess Parvati who is worshipped as Girija Mata. The rock formation in the form of a Shiva linga is considered to be a manifestation of Lord Shiva.

So far, no one has climbed Manimahesh Kailas, in spite of the fact that much taller peaks have been scaled, including Mount Everest. According to one legend, a local tribesman tried to climb it once along with his herd of sheep and was turned into stone along with his flock. The minor peaks around the main peak are believed to be the remnants of the shepherd and his sheeps.[10]

KINNAUR KAILAS is a mountain in Himachal Pradesh. It is considered sacred to both Hindus and Buddhists. According to legend, the shrine of Kinnaur Kailas was present since the time of the demon Bhasmasura, who had received a blessing from Lord Shiva that anyone who touched his head would be turned to ashes. Bhasmasura, seeing the powerful effect of this boon, tried to destroy Lord Shiva himself. Lord Shiva hid himself in Kinnaur Kailas in order to save himself. Finally, it was Lord Vishnu who killed the demon. The Shiva linga of

Kinnaur Kailas is a 79-ft vertical rock. On a clear day, one can see the Shiva linga, which changes colour during the course of the day. This is a very sacred place as it is associated with Lord Shiva and Goddess Parvati. A natural pond or kund near Kinnaur Kailas peak, known as Parvati Kund, is considered to be a creation of Goddess Parvati. It was a meeting place for the two. Lord Shiva conducts a meeting of goddesses and gods at Kinnaur Kailas every winter, which accounts for the yatra that takes place every year at this time.[11]

Haryana

DHOSI HILL is the most sacred hill in Haryana. It is an extinct volcano at the north-western end of the Aravalli mountain range, with temples, a pond, fort, caves and a forest. The hill has all the features of a perfect volcanic hill with a crater: solidified lava is still visible on one side of the hill and there is a perfect conical view from top. According to the Mahabharata, the volcano erupted in the beginning of the treta yuga. It was described by Guru Shaunaka, who had accompanied the Pandava brothers during their visit to the hill. The Mahabharata also describes how the hill has been revered since then because it was inhabited by respected rishis and munis. Halfway up the hill from the south side is a reservoir known as Shiv Kund. This is filled from the reservoir at the summit. Apart from temples at Shiv Kund, there are several more on the crater. Most prominent among them is the Chyavana Rishi Temple. Different temples attract devotees on different days. On Somvati Amavasya day, people assemble on the hill for a holy bath in the sarovar. Those visiting the hill on pilgrimage perform a parikrama (circumambulation). The 8–9-km parikrama track includes some portions damaged by landslides. There is a cave on the route where Chyavana performed tapas for years, and it now provides shelter to pilgrims.[12]

Uttarakhand

BANDARPOONCH means 'monkey's tail' in Hindi and is situated in the western Garhwal region of Uttarakhand. According to legend, when Hanuman's tail caught fire before the battle to rescue Sita from the demon Ravana in Lanka, he went to the summit of this hill to extinguish it.[13]

CHAUKHAMBA MOUNTAIN is a ridge that is part of the Gangotri group and the Garhwal Himalayas. The peak is in the holy city of Badrinath and lies at the head of the sacred Gangotri glacier.

HAATHI PARVAT, situated in the Chamoli district in the Garhwal Himalayas in Uttarakhand, means 'elephant mountain', a reference to its similarity to the figure of an elephant. Two huge rocks on Haathi Parvat are described as a crow and an eagle. It is said that a learned Brahmin of Ayodhya once incurred the wrath of sage Lomas who lived here and was changed into the form of a crow by the sage. The summit is sacred because River Dhauliganga flows nearby.[14]

CHANDRASHILA means 'moon rock' in Hindi. It derives its name from an ancient legend that the moon god Chandra once did penance here. Some also believe that Lord Rama did penance here after defeating Ravana.

DUNAGIRI OR DRONAGIRI PEAK in the Kumaon region of Uttarakhand is known for the Shakti temple known as Dunagiri Devi temple. Many saints and savants have visited this hill, including Garg Muni and Sukhdev Muni. River Gagas is named after Garg Muni. Sukh Devi also belonged to this region, where the ashram of Sukhdev Muni was located. When Lakshmana was hit by Indrajit's arrow, Hanuman brought the Sanjivani herb from Dronagiri Mountain to save his life. The Pandavas stayed here for some time during their travels. Guru Dronacharya, one of the Pandava gurus, also devoted his time to practise his *tapasya* at this place. In the eighth to ninth centuries CE, Adi Shankaracharya moved here to begin monastic orders.

Dunagiri is mentioned in the *Skanda Purana*. Dunagiri Devi is described as Mahamaya Harpriya ('Manaskhand', 36.17–18). 'Manaskhand' of *Skanda Purana* bestows Dunagiri with the title of Brahma Parvat (Divine Mountain). Among all the Shakti temples of Kumaon, Dunagiri is believed to be the most ancient Siddha Shaktipeeth. It was influenced by Shaiva, Vaishnava and Shakta practices.[15]

GANGOTRI is situated in the Uttarkashi district of Uttarakhand state, on the banks of River Bhagirathi. It is one of the Char Dhams of Uttarakhand. According to Hindu mythology, Ganga was transformed into a river at this spot in order to release the ancestors of King Bhagirath from their sins. The king meditated here to bring Ganga down from heaven to earth. Hence the river is also called Bhagirathi. There is a submerged Shiva linga here, which is the spot where Lord Shiva held Ganga in his hair, when she descended from heaven and became a river. The Shiva linga remains underwater during summer and is only visible during winter, when the water level drops. The snow leopard, black bear, brown bear, Himalayan tahr, Himalayan monal, Himalayan snowcock, bharal (the famous blue sheep of Himalayas), ibex and many other mammals and bird species are found at the Gangotri National Park.

NANDA KOT lies in the Kumaon Himalayas, just outside of the ring of peaks enclosing the Nanda Devi Sanctuary, 15 km south-east of Nanda Devi. The name Nanda Kot literally means 'Nanda's fortress' and refers to Goddess Parvati who made her home among the ring of mountains in the region.

NEELKANTH HILL stands today at a place where it is believed there was no mountain before. Earlier, there was a route between Kedarnath and Badrinath. The *purohita* of the two temples worshipped them in one day. Due to the sins of the worshiper, Lord Shiva became displeased with him and blocked the way using a huge mountain, which is the modern Neelkanth.

OM PARVAT is also known as Little Kailas. The pattern of the snow in the form of the Om (Aum) symbol on this peak holds great significance in Hinduism, Buddhism and Jainism.

PURNAGIRI OR PUNYAGIRI MOUNTAIN, known as the 'mountain of good deeds', is one of the 108 Siddha Peeths, with a temple dedicated to Goddess Purnagiri. The River Kali begins its descent into the plains from this hill.

SWARGAROHINI forms the path to heaven taken by the Pandavas, Draupadi and their dog. It is believed that this is the only way to reach heaven in the human form.

TRISHUL consists of three mountain peaks resembling a trident, the symbol of Lord Shiva.

YAMUNOTRI is the source of the River Yamuna and the seat of Goddess Yamuna. It is one of the four sites in India's Chhota Char Dham pilgrimage. The Yamuna, like the Ganga, is a divinity in Hinduism.

Eastern India has several mountain ranges with peaks that are regarded as sacred.

Arunachal Pradesh

TAKPA SHIRI is a peak that remains unclimbed because it is one of the most sacred mountains here. It lies at an altitude of 6654 m in the state of Arunachal Pradesh, on the border between China and India. Locals believe that Takpa Shiri confers spiritual bliss. Circumambulating it is considered as sacred as circumambulating Mount Kailas.[16]

Assam

BAGHESWARI HILL is situated 1 km away from Bongaigaon town in Assam. The hill houses an ancient Shiva temple inside a stone cave flanked by two other temples—Bagheswari and Baba Taraknath—on either side. Bagheswari Temple is one of the oldest and most visited temples of Assam, dedicated to Goddess Durga. It is one of the fifty-one Shakti Peeths. According to Hindu mythology, this is the place where Goddess Durga's trishul (trident) fell when her body was cut into pieces by God Vishnu, after Daksha's yajna.[17]

CHITRACHAL HILL, also known as Navagraha Hill in Guwahati, Assam, houses the famous temple of nine planets. In this temple are nine Shiva lingas, each adorned with a coloured garment symbolic of each of the planets, with a

Shiva linga in the centre symbolizing the sun. Assam was said to have been a great centre of the study of astronomy and astrology, which is one of the reasons why it was called Pragjyotishapura or the City of Eastern Astrology. According to the *Kalika Purana*, Pragjyotishapura was created by Brahma as he wanted to build a city equivalent to Indra's heaven.[18]

GANDHAMOAN HILLS near Hajo are famous for the Shiva temple or *devalaya* where Lord Shiva is worshipped as Bhringeswar or Lord of Bhringi, his devotee. The main object of worship here is a big turtle-shaped stone representing Lord Shiva and which remains submerged in the water tank for five to six months. The most special feature of this temple is that animal sacrifices are offered here.[19]

HATIMURA HILL is on the north bank of the River Brahmaputra, about 40 km west of Guwahati. On the western slope of this hill stands Dhareshwar Devalaya, one of Kamrup's most ancient and historically famous shrines, said to have been built by the Ahom king Shiva Singha in about 1730 CE. The origin of the Dhareswar Shiva Sthana goes back to pre-Vedic times, so believe the local people. The place in and around the sacred site was originally inhabited by the Kacharis. According to legend, a cow owned by a Kachari always went missing at a certain place up the Hatimura Hill. After investigation, the villagers learnt that the cow used to offer her milk at a particular place on the hill. The curious villagers dug up this place and discovered a stone Shiva linga, which they started to worship as Dhareshwar Shiva. The devalaya is located in beautiful natural surroundings. Water from a stream up on the hill flows to the sacred site of the linga.[20]

NEELACHAL OR KAMAGIRI HILL, situated 7 km from Guwahati, is of great historical, archaeological and religious importance. This hill was believed to be a Khasi sacrificial site earlier; now there is a group of ancient temples on the top, the most famous being the Kamakhya Temple. It is one of the oldest and most revered centres for the worship of Shakti and the tantric cult.

It is believed that the female genitalia (*yoni*) of Sati fell here while her corpse was being carried by her husband Shiva, turning the hill blue; hence the name Neelachal (blue hill). There is no image of Shakti here. In a corner of a cave in the temple, there is a sculptured image of the yoni of the goddess, which is the object of reverence. A natural spring keeps the stone permanently moist. Other temples on the Neelachal Hill include those of Tara, Bhairavi, Bhuvaneswari and Ghantakarna.

MONIKUT HILL is situated in Hajo in Assam, about 30 km west of Guwahati. The famous Hayagriva Madhava Temple is situated on this hill. The present temple was constructed by King Raghudeva Narayan in 1583. Some Buddhists believe that the Hayagriva Madhava Temple is where the Buddha attained nirvana. The presiding deity is Narasimha, an incarnation of Vishnu. There is a big pond known as Madhab Pukhuri near the temple where Doul, Bihu and Janmashtami festivals are celebrated. This temple is revered by both Hindus and Buddhists, attracting Buddhist monks from everywhere.[21]

SANDHYACHAL HILL is located in Beltola in Guwahati, Assam, where the eighteenth-century Basistha (Vashishtha) Temple, also known as Basistha Ashram, is located, believed to be the last monument built by the Ahom kings, while the Puranas connect this temple to the legendary sage Vashishtha. It is said that the hillock was named Sandhyachal by the devas, for Vashishtha performed *sandhya* (meditation) on this hill. The *Kalika Purana* has described the Basistha Temple as one of the seven Shakti Peeths. The Shiva-Shakti Peeth, also known as Tara Peeth, is a very sacred part of this temple. The waters of the three streams—Sandhya, Lalita and Kanta—that flow here are said to have medicinal value and lead to longevity.[22]

Manipur

KAINA HILL is situated about 14 km east of Manipur valley. The Meitei Vaishnavas call it Bhashmukh Parbat. According to popular legend, Lord Krishna appeared to his devotee Bhagyachandra, the maharaja of Manipur, in a dream and asked him to build a temple with his image carved out of a jackfruit tree at Kaina. Beautiful hill shrubs and charming natural surroundings give the place a saintly solemnity. Rasaleela dances depicting the divine dream are performed here.[23]

Meghalaya

JAINTIA HILLS are located in the state of Meghalaya. The Marangksih Peak on the eastern plateau of Jaintia Hills, standing majestically at a height of 1631 m from the mean sea level, is the highest peak in the entire district. The Jaintia Hills are richly endowed with natural resources. Nartiang, 65 km from Shillong, was the summer capital of the Jaintia kings. Monoliths exist throughout the length and breadth of the Khasi and Jaintia Hills. However, the biggest collection of

monoliths or megalithic stones in one single area is to be found
north of the Nartiang market. These consist of menhirs and
dolmens locally known as Moo Kynthai. The tallest menhir
was erected by U Mar Phalyngki, a trusted lieutenant of the
Jaintia king, to commemorate his victory in battle. Other
monoliths were erected by U Mar Phalyngki, U Luh Lyngskor
Lamare and various clans of Nartiang village between 1500
and 1835 CE. The monoliths represent the megalithic culture of
the Hynniewtrep people. A 500-year-old temple of Goddess
Durga is another attraction at Nartiang. In addition to these,
the hills have numerous sacred lakes and caves which are
revered by the local people.[24]

Sikkim

MOUNT KANCHENJUNGA, the third highest peak in the world, is
situated in the Kanchenjunga National Park, bordering Nepal,
in Sikkim. On its lower slope is a wet temperate forest which
gradually rises to bare rock, ice and snow. It is bound on the
west by the River Tamur and on the east by the River Teesta.
The mountain has five peaks, four of them over 8450 m, of
which two are in Nepal. It is traditionally worshipped by the
people of Sikkim and Tibet. Kanchenjunga literally means 'the
five treasures of snow', which represent the five repositories of
god, namely gold, silver, gems, grain and the holy book. In
Tibetan, the peak is known as Demoshang. Kanchenjunga is
represented by a fiery red-faced deity, a crown of five skulls
and a snow lion. The area around the peak is reputed to be the
home of the yeti, also known as the Kanchenjunga demon.[25]

SAMDRUPTSE HILL, about 75 km from Gangtok, is a renowned
pilgrimage centre of the Buddhists. The word Samdruptse

literally means 'wish-fulfilling tree' in the Bhutia language. It is considered to be a dormant volcano. Myths say that the Buddhist monks have been going to the top of the hill and offering prayers to it to keep it calm. Samdruptse gained prominence mainly due to a gigantic statue of Guru Padmasambhava, also known as Guru Rinpoche. On the hillock behind the statue, Buddhists place prayer flags and build cairns in order to obtain good luck from this hill.[26]

TENDONG HILL in the Lepcha language means 'the upraised horn'. According to a legend of the Lepcha tribe, there was once a heavy downpour for forty days and forty nights and consequently, a great deluge. The entire tribe was in great danger of being drowned. Miraculously, the hill rose and the people were able to ascend the peak. They were thus saved. From that time onwards, the hill has been held in great reverence by the Lepchas and is regularly worshiped.[27]

Tripura

UNAKOTI HILL lies about 178 km north-east of Agartala in Tripura. The rock-cut carvings, belonging to the eleventh and twelfth centuries CE, are the largest bas-relief sculptures in India. It is virtually an open-air art gallery, besides being a Shaiva pilgrimage site dating back to the seventh, eighth and ninth centuries CE.[28]

West Bengal

DURPIN DARA is one of the two hills on which the town of Kalimpong in West Bengal stands. The hill commands a panoramic view of Kalimpong, the Himalayas, the River

Teesta and the Jelep La. A famous monastery called Zang
Dhok Palri was built by Dudjom Rinpoche on the hilltop
in 1946 and consecrated by the Dalai Lama. The monastery
holds 108 volumes of the Kangyur as well as other holy books
and scrolls that were moved out of Tibet after the Chinese
invasion.[29]

AYODHYA HILL, located about 42 km from Purulia, is a part of
the Dalma Hill Range between West Bengal and Jharkhand.
It has dense forests teeming with elephants and other wildlife.
According to legend, Rama and Sita stayed here during their
exile. On one occasion, when Sita felt thirsty and asked Rama
to quench her thirst, he shot an arrow into the earth and out
gushed a stream of water. This area is known as Sita Kund.
People belonging to the Tundra community drink water
from here before they go hunting as it is considered sacred. A
beautiful Ram Mandir was constructed 500 m away from the
Sita Kund, and prayers are offered at both sites.[30]

MAMA BHAGNE PAHAR is also known as Pahadeshwar (god
of the rock) or Bongaburu, according to the Santhal tribes.
There are two legends associated with this hill. According
to the first, when Lord Rama decided to wage war against
Ravana, he had to build a bridge across the sea. He went
to the Himalayas to collect stones for building the bridge.
As he was returning south, his horses took fright and some
of the stones fell out of his chariot, and gave rise to the
hill. The other story says that the stones were collected by
Vishvakarma at the command of Lord Shiva who wanted to
build a second Kashi. He was about to commence the work
when daylight dawned and he left. The stones left behind
formed the sacred rock.[31]

Odisha

NIYAMGIRI HILLS, in the districts of Kalahandi and Rayagada, are sacred to the Dongria Kondh tribe. Though forest clearance had been given to mine bauxite here, the tribes agitated against the project, following which the Hon'ble Supreme Court permitted the tribes to decide whether they wanted the project or not. They rejected it, and the Ministry of Environment and Forests scrapped the project. Their contention was that it would destroy local biodiversity and pollute the two rivers and the thirty-six streams that come down from the hill and irrigate their crops.[32]

Bihar

BARABAR HILLS, consisting of four ancient caves called Karan Chaupar, Lomas Rishi, Sudama and Visva Zopri, are among the most ancient sites of rock-cut architecture in India. They were constructed around 200 BCE by the great Mauryan king Ashoka, for the use of the Ajivika sages. The hills go back to the Mahabharata period. They are named after Vanasura, the *senapati* or general of King Jarasandha, who constructed a fort on the hills, the remains of which can be seen even today. Ashvatthama, the son of Drona and the murderer of the Pandava children, is said to be wandering even today in the valleys of Barabar. He became immortal by the grace of Lord Shiva, whose temple is situated on top of the highest peak of Barabar Hills.

MANDARA PARVAT is a small mountain in Banka district, also described as Sumeru Parvat. The mountain was the churning pole of the samudra manthana (churning of the nectar of immortality). Due to this association, the hill is a site of great

religious significance. The patterns on the rock are believed to
have been made by the great snake Vasuki who was used as a rope
by the devas and asuras when they churned the ocean of milk. An
inscription of the Gupta king Adityasena discovered on Mandara
Hill says that he and his queen Sri Konda Devi had installed an
image of Narahari (Narasimha) on the hill and that the queen had
excavated the Papa Harini (or Manohar Kund) tank at the foot of
the hill. The twelfth Jain Tirthankara Vasupujya attained nirvana
here and hence there is a Jain temple on the hill.[33]

Jharkhand

KOLHUA HILL in the Chatra district is considered to be a Siddha
Peeth as it is believed that the womb (*kokh*) of Goddess Sati
fell here. Hindu devotees come here to pray for children. Lord
Rama, Sita and Lakshmana, it is believed, stayed here for a
while in exile. It is also believed that the Pandavas stayed here
during their *agyatvaas* (period in disguise).[34]

PARASNATH HILL is an important pilgrimage centre for the Jains
who call it Sammet Sikhar. The peak is named after Parasvanath,
the 23rd Tirthankara of the Jains. Parasvanath is also known as
Marang Buru or hill deity of the Santhal tribes. The pilgrimage
consists of a circumambulation on foot for 30 km through the
Madhuban forest. Along the track are shrines to each of the
twenty-four Tirthankaras. Recently, the site was declared a
wildlife sanctuary.[35]

Uttar Pradesh

GOVARDHAN HILL, situated in Mathura district, is a famous
pilgrimage site. A small sandstone hillock covered with many

flowering plants and shaped in the form of a peacock, it is also known as Giriraj. The hill derives its importance from the legend of Lord Krishna, who is believed to have lifted the hill with his little finger and held it up for seven days and seven nights in order to protect the people of Vrindavan from the heavy rains caused by Lord Indra.

KAMADGIRI is a mountain and a popular holy site in Chitrakoot Dham, which is believed to have been the abode of Rama, Sita and Lakshmana during their exile. The association with the Ramayana is a unique spiritual legacy enjoyed by the town. A forested hill, Kamadgiri, venerated today as the embodiment of Rama, is skirted all along its base by a chain of temples. It has a pedestrian path at the base, which is used by barefoot pilgrims for circumambulation. There are several temples on the 5-km pathway of the parikrama, one of which is the famous Bharat Milap Temple where Bharata met Lord Rama.[36]

Rajasthan

MOUNT ABU is Rajasthan's only hill station, situated at the southern tip of the Aravalli Range. The highest peak on the mountain is Gurushikhar, which is the highest point between the Nilgiris in the south and the Himalayas in the north. As per legend, Mount Abu came into existence when Nandini, sage Vashishtha's wish-fulfilling cow, fell into a deep lake. The sage appealed to the gods, who sent Arbuda, the celestial cobra. Arbuda carried a huge rock on his head and dropped it into the lake, displacing the water. Thus, Nandini was saved. The place came to be known as Arbudachala (the hill of Arbuda) after the mighty serpent.

The area is rich in floral biodiversity and subtropical thorn forests. The latter, found in the foothills, gradually give way to subtropical evergreen forests at higher altitudes. Besides over 800 species of plants, Mount Abu is the only place in Rajasthan, a desert state, where one can find orchids. It is home to several species of fauna including the endangered sloth bear. It is also the location of the Dilwara temples, a complex of Jain temples.

Gujarat

ARASUR HILL is located in the Banaskantha district, where the famous temple of Ambaji dedicated to Mother Goddess is located. It is a Shakti Peeth, for it is believed that the heart of the goddess fell here. The temple does not contain any idol, only a *yantra* engraved in a niche.

CHOTILA MOUNTAIN, near Rajkot, is the site of the temple of Mata Chamunda, one of the sixty-four avatars of Shakti and the *kuladevi* (family goddess) of most Hindus of Saurashtra.

GABBAR HILL is one of the fifty-one Shakti Peeths in India. Many believe that the heart of Devi Sati fell here. It is the abode of Goddess Ambaji, an ancient deity whose footsteps are imprinted on the hill. According to the Ramayana, Lord Rama and Lakshmana came to the ashram of Rishi Shringa in search of Sita. They were advised to worship Devi Ambaji at Gabbar. Rama did so and the goddess gave him a miraculous arrow named Ajay, with the help of which Rama conquered and killed his enemy Ravana in the war.[37]

Madhya Pradesh and Chhattisgarh

There are many hills in Madhya Pradesh sacred to the Hindus and Jains. They include Bharveli Mountain where the twin temples of Gomji-Somji and Jwaladevi, dedicated to Goddess Parvati, are situated; the Chauragarh Hill with the shrine for Lord Shiva on top; the Gopachal Hill with 1500 Jain idols; the Muktagiri Hill with its Jain shrines; the Sanchi Hill with its excellently preserved stupas, the earliest dating to 300 BCE; and others.

AMARKANTAK is the source of the Narmada and Son rivers. It is a beautiful forested area situated at the meeting place of the Vindhya and Satpura Ranges. Amarkantak is sacred to Hindus and is deemed to be the gateway to nirvana. There are several temples in and around the region, including an ancient temple dedicated to Goddess Narmada, the temple of Devi Sati and the Trimukhi Temple dedicated to Lord Shiva, which is the oldest temple in Amarkantak.

DONGARGARH HILLS are the site of the famous Maa Bamleshwari Temple. *Dongar* means mountains while *garh* means fort. The temple is a major spiritual centre of Chhattisgarh.

GADIYA MOUNTAIN in Kanker is well-known for a tank that never dries up, even in the hot summer season. There is a Jogi cave, which is 50 m in length, where saints used to live and meditate in the past. The cave could accommodate 500 people and possesses a small pond, the water from which descends on to the rocks below in a spectacular waterfall. River Doodh flows by the foot of the mountain.

The Vindhyas

Seeing the sun, moon and stars circumambulating Mount Meru, the Vindhya Mountain wanted a similar honour. 'Why do you do a pradakshina around the Meru alone?' he asked the sun. 'I can't help it,' replied Surya. 'My path was fixed by those who created the world.' Not satisfied, Vindhya grew and grew in order to obstruct the path of the heavenly objects. The devas then approached Agastya and requested him to stop the mountain from growing further. In the interest of the world, Agastya went to Vindhya and asked the mountain to bend low, as he wanted to go south, and to remain in that position till he returned. Vindhya agreed. Agastya, however, stayed in the south and the Vindhyas have remained a low mountain range since then.[38]

The Vindhya Range divides north and south India. According to the 'Kishkindha Kanda' of the Valmiki Ramayana, the architect Maya built a palace in the Vindhyas. When Rama entered Dandakaranya, the Vindhya region, he was warned about the unknown territory infested with demons. The Vindhyas are the residence of Vindhyavasini, a form of Shakti who lived there after killing the demon. There is a temple dedicated to her in Vindhyachal, a town in Uttar Pradesh.[39] The Mahabharata describes the Vindhyas as the abode of Kali.

The Vindhyas do not form a single mountain range; they cover several hills, including the Satpura range. They run north of and parallel to the River Narmada in Madhya Pradesh and extend up to Gujarat in the west and Bihar in the east. Many ancient texts mention the Vindhyas as the southern boundary of Aryavarta. Several rivers originate from these hills, making them ecologically significant.

Western Ghats or Sahyadri

Beginning at the border between Gujarat and Maharashtra, the Western Ghats is a mountain range that runs parallel to the west coast of India and ends in Kanyakumari in south India. It consists of national parks, wildlife sanctuaries and reserve forests that have been designated as world heritage sites—four in Maharashtra, ten in Karnataka, five in Tamil Nadu and twenty in Kerala. The hills are among the world's ten biodiversity hotspots. They form one of the four main watersheds of India, with the Rivers Godavari, Krishna, Kaveri, Tungabhadra and Tamraparani originating here, apart from the many streams draining the hills that join the rivers, carrying large volumes of water into the Bay of Bengal. They are the home of the largest number of sacred groves in India and rare tribal and folk deities like Hulideva in Karnataka, Vaghdeo in Maharashtra—both tiger gods—and the crocodile deity in Goa.

Maharashtra

With the Western Ghats running through the state, there are many sacred hills, most of which have sacred groves and temples of local tribes.

MAHABALESHWAR in the Satara district is a popular hill station known for its ancient temples. The most important is the Pancha Ganga Temple believed to be the site of the origin of five rivers, Venna, Gayatri, Savitri, Krishna and Koyna. Mahabaleshwar derives its name from Lord Mahabali or Shiva who appears as a swayambhu linga in the shape of a *rudraksha* seed.

RAMTEK HILL was the place where Rama rested while he was in exile. The ashram of the great sage Agastya was situated close to the hill. When the sages performed religious rites, demons used to trouble and kill them. When Lord Rama heard about this, he vowed to destroy the world of demons. *Tek* means vow, hence Ramtek means vow of Rama. It is believed that whoever takes a vow at Ramtek is blessed by the gods for its fulfilment.

Karnataka

CHAMUNDI HILLS is famous for the temple of Chamundi or Durga, the slayer of Chanda, Munda and Mahisha. She is the presiding deity of ancient Mysore (now Mysuru). At the foot of the hill is an enormous statue of Mahisha, the demon killed by the goddess. Mahisha was once the ruler of Mysore, which is still named after him—Mysore/Mysuru is derived from Mahisha-uru (the land of Mahisha). The *Skanda Purana* and other ancient texts mention the Trimuta Kshetra surrounded by eight hills. Chamundi Hill, one of the eight, is on the west. The hill has played an important role in the ecology and climate of Mysuru city. They are a repository of biodiversity and also act as a watershed and a source of groundwater recharging. The maharajas of Mysuru developed several lakes and dug tanks and ponds, all of which act as the main water sources for the surrounding villages.

HEMAKUTA HILL is located on the southern side of the Hampi village in Karnataka. According to mythology, Lord Shiva performed penance on Hemakuta Hill before marrying Pampa, the river below the hill. Due to this decision of Lord Shiva, there was a rain of *hema* (gold) on the hill, hence the name

Hemakuta. Kishkinda, where Lord Rama met Hanuman, Sugriva and the other Vanaras, is situated across the river from Hemakuta.

BILIGIRI–RANGANA HILLS is situated in the Biligiri-Ranganathaswamy Temple Wildlife Sanctuary. It is at the confluence of the Western and Eastern Ghats, and has been declared a tiger reserve. Biligiri means white rock in Kannada, and it is here that the temple of Lord Ranganatha is situated. The hills are covered with rich deciduous forests that are home to tigers, gaur and the Asian elephant, apart from other mammals and over 250 species of birds.

MATANGA HILL on the River Tungabhadra has many associated legends. The hill was the hermitage of sage Matanga. According to a legend, the monkey prince Vali killed a buffalo demon called Dundhubi and threw the corpse on the sacred hill. Angry at this act, sage Matanga cursed Vali that he could never step on the hill. Later, Dundhubi's son Mayavi fought with Vali to avenge his father's death. Vali chased him into a cave and asked his brother Sugriva to stand guard outside. After a while, thinking that Vali had been killed in the fight, Sugriva closed the mouth of the cave. Finally Vali emerged from the cave and chased his brother out of the kingdom. Sugriva and his general Hanuman took refuge in Matanga Hill, as Vali could not climb it. Later, Lord Rama killed Vali and crowned Sugriva the king of the Vanaras.

CHANDRAGIRI AND VINDHYAGIRI HILLS are situated in Shravanabelagola. These hills are among the oldest Jain pilgrim centres in the south. The tombs of Chandragupta Maurya, Bhadrabahu Muni and many other great Jain devotees are situated here. Chandragiri is named after Emperor Chandragupta Maurya, who, in 300 BCE, became a Jain ascetic and moved south with his guru Acharya Bhadrabahu. The famous 58-ft-high statue of Gommateshvara, the world's largest monolithic statue, is located on Vindhyagiri Hill. Every twelve years, thousands of devotees congregate here to perform the *mahamashtakabhisheka*, a ceremony in which the statue is anointed with water, turmeric, rice flour, sugar-cane juice, sandalwood paste, saffron and gold and silver flowers. Chandragiri has fourteen Jain *basadi*s of which the first, Chandragupta Basadi, was constructed by Emperor Ashoka in the third century BCE in memory of his grandfather Chandragupta Maurya.

Goa

CHANDRANATH HILL is the watershed of River Paroda. The ancient Chandreshwar Temple, also known as the Chandranath Temple, is situated about 14 km from Margao on the Chandranath Hill. The unique feature of the temple is that on a full moon night, water oozes out of the Shiva linga, carved out of a rock, whenever moonlight falls on it. The temple is so designed that moonlight falls on it on every full moon. The location, topography, antiquity and natural beauty of the hill are unique. According to an ancient Sanskrit inscription, the temple has stood on this magical spot for nearly 2500 years. The shrine has a cavernous inner sanctum, hollowed from a huge boulder, around which a typical Goan-style structure, capped with a red-tile room and domed sanctuary, was constructed in the seventeenth century. The temple is associated with the Bhoja dynasty, who were the rulers of the region till the eighth century. Their family deity was Lord Chandreshwar and their capital Chandrapur (Chandor) was named after the deity.

Eastern Ghats

Most of the Eastern Ghats run through the Mahabubnagar and Nalgonda districts of the state of Telangana and the Kurnool, Guntur, Prakasam, Kadapa and Chittoor districts of the state of Andhra Pradesh. They run in a north–south alignment, parallel to the Coromandel Coast and between the Rivers Krishna and Pennar. The northern boundary is marked by the flat Palnadu Basin while in the south it merges with the Tirupati Hills. An extremely old system, the hills have extensively weathered and eroded over the years.

Andhra Pradesh

SIMHACHALAM OR SIMHAGIRI, located in Visakhapatnam, is known for the eleventh-century temple dedicated to Lord Varaha Narasimha. The forests in the hills of Simhachalam are known for their rich biodiversity. They were once hospitable grounds for black panthers and also the mouse deer. The hill range is a part of the Eastern Ghats and is called Kailas.

TIRUMALA HILL (also known as Saptagiri in Sanskrit) comprises seven peaks—Seshadri, Neeladri, Garudadri, Anjanadri, Vrishabhadri, Narayanadri and Venkatadri (Figure 6.5). The famous temple of Lord Venkateshwara is located on the seventh peak, Venkatadri. Tirumala is supposed to be a part of the sacred Mount Meru. According to legend, a contest once arose between the god of wind, Vayu, and Adisesha, the divine

serpent couch of
Vishnu. Lord Vayu
tried to blow out all
the peaks of Meru,
while the serpent
tried to protect
them with his
hoods. After some
time, Lord Vayu

Figure 6.5: Tirumala Hill, Tirupati,
Andhra Pradesh

Source: The C.P. Ramaswami Aiyar Foundation's photo archives

was exhausted and discontinued his blowing. Thinking that he
had won the contest, Adisesha lowered his hood, when Lord
Vayu blew at the peak. One part of the peak fell at Tirupati,
forming the sacred Tirumala Hill.

Tirumala Hill is covered with lush green forests. It is a
part of Sri Venkateswara National Park, where wild animals
roam freely. The area is also rich in medicinal plants and herbs.
The hill range is dotted with several waterfalls such as Jabali
Teertham, Rama Teertham, Sesha Teertham, Bala Teertham,
Papavinasam and Akash Ganga.

NALLAMALA HILLS are thickly forested and remain largely
unexplored. Srisailam on River Krishna is known for an
ancient temple of Lord Shiva and the Ahobilam Nava
Narasimha Temple. The famous Srisailam Tiger Reserve is also
situated here. A stream flows down the peak and culminates
in a beautiful waterfall and sacred pond called Namaligundam
(peacock pool). Ahobilam, located about 150 km from
Kurnool, is a famous Vaishnava kshetra located on Nallamalla
Hills. This is the place where Lord Vishnu incarnated as Lord
Narasimha to save his devotee Prahlada from his evil demon
father, Hiranyakashipu. He was eventually killed by Narasimha
on top of the hill. *Bilam* means cave in Telugu. Some of the

temples are located inside caves or on top of the mountain, hence the name Ahobilam. Later, goes the legend, Narasimha married a local Chenchu tribal girl belonging to this region.

Tamil Nadu

PALANI HILLS near Madurai are the location of the temple of Murugan or Karttikeya, son of Lord Shiva. Sage Narada once visited Lord Shiva at Mount Kailas and presented the *jnana palam* (fruit of knowledge) that held the elixir of wisdom. Shiva decided to award it to one of his two sons—Ganesha and Karttikeya—who first circled the world three times. Lord Karttikeya started off on his peacock mount. However, Ganesha, who understood that the world was contained in his parents, Shiva and Parvati, circumambulated them thrice. Pleased with their son's wisdom, Lord Shiva awarded the fruit to Lord Ganesha. When Karttikeya returned, he was furious to learn that his brother had won. He left Kailas for the Palani Hills in south India and chose to remain as a hermit, discarding his grand robes and ornaments. The temple is situated on Sivagiri Hill.

Kodaikanal, situated in the Palani Hills, suffered mercury poisoning of its groundwater, a proven case of mercury contamination caused by a multinational company in the process of making mercury thermometers for export. The exposure of the environmental abuse led to the closure of the factory in 2001. As much as 290 tonnes of mercury waste had been dumped in the forest, but the company was forced by the local people to send it back to the United States for recycling in 2003.[40]

Of the six sacred abodes of Karttikeya in Tamil Nadu, four are situated on hills: Palani, Tiruttani, Tirupparakunram and

Palamudircholai. In Tamil literature, Murugan is described as the god of the kurinji or hill region. Tiruttani is near Tirupati while Tirupparakunram and Palamudircholai are situated near Madurai.

TIRUVANNAMALAI HILL is the site of the temple of the same name (Figure 6.6). Once, Shiva appeared as a flame and challenged Brahma and Vishnu to find his source. This form of Shiva is called Lingodbhava. Brahma became a swan and flew into the sky to see the top of the flame, while Vishnu became the boar Varaha and searched for its base. Neither Brahma nor Vishnu

Figure 6.6: Tiruvannamalai Hill, Tamil Nadu

Source: The C.P.R. Environmental Education Centre, www.cpreecenvis.nic.in

could find the source. Vishnu conceded defeat, but Brahma lied and said that he had found the pinnacle. As punishment, Shiva ordained that Brahma would never have temples on earth for his worship.[41] This incident is believed to have happened on this hill. Of the five *panchabhuta linga*s, the Tiruvannamalai linga represents agni or fire. Tiruvannamalai is named after the main deity of the temple, Lord Annamalai. During the festival of Kartigai, 1,00,000 oil lamps are lit on the Annamalai Hill. The event is witnessed by three million pilgrims. On the day preceding each full moon, pilgrims circumambulate the base of the hill in a practice called *girivalam* (circumambulating the hill). Tiruvannamalai was also the location of the ashram of Swami Ramana Maharshi, a great Hindu saint of the twentieth century.

NILGIRI MOUNTAINS are worshipped by the Toda tribe, a pastoral community who preserve the grasslands and the shola forests on the upper plateau of the Nilgiris. Mountains are their temples and places of worship, with each god having his own separate hill where he dwells. The mountain Porhsthitt, near Thalaptheri Mundh, located on the western part of the Nilgiris upper plateau, is worshiped by the Todas even today. There are two important reasons for worshipping the mountains. Firstly, the grasslands are necessary for the buffaloes; secondly, the Todas strongly believe that the hills, such as Mukurti (Figure 6.7), are the dwelling places of the gods. The hill Kotran Dekarsh

Figure 6.7: Mukurti Peak, Nilgiris, Tamil Nadu

Source: The C.P.R. Environmental Education Centre, www.cpreecenvis.nic.in

(Malleshwaran Hill) is located on the southern side of the Nilgiris, on the Attappady Valley. This steep hill is associated with Lord Shiva.

Kerala

SABARIMALA or MOUNT SABARI is the abode of Lord Ayyappa, situated in the midst of eighteen hills in the Western Ghats. The hill is named after an old woman, Sabari, who attained salvation after meeting Lord Rama. The temple is surrounded by mountains and dense forests of the Periyar Tiger Reserve. It is believed that the deity of the temple was consecrated by Lord Parashurama at the foot of the Sabari Hills. According to legend, this is the location where Lord Ayyappa meditated after annihilating the dreaded demon Mahishi. It is one of the largest pilgrimage centres in the world, which is posing a threat to the tiger reserve.

PONNAMBALAMEDU HILL is an ecologically sensitive area and a critical habitat for tigers. The name is derived from the Malayalam words pon (gold), *ambalam* (temple) and *medu* (hill). Thus Ponnambalamedu means 'hill of the golden temple'. It is the scene of the *makaravilakku* event which is conducted in the presence of half a million pilgrims annually. Makaravilakku refers to the appearance of the holy flame atop the mountain during the annual festival of Sabarimala. It marks the final phase of the pilgrimage season of Sabarimala which lasts about two months.

MADAYIPARA HILL is the site of the Madayikavu (Thiruvar Kaadu) Bhagavati Temple dedicated to Goddess Kali. The Vadukunnu Temple, also situated on the plateau, is dedicated to Lord Shiva, but was destroyed by the army of Tipu Sultan in the eighteenth century. It has since been rebuilt.

7

Conclusion

During every harvest, after the grain has been cut, the women collect the seeds and preserve them in a storage bin made of baked clay. They decide on the amount of seeds to be preserved, the varieties and the method of preservation. The terracotta bin is sacred and a lamp is lit beside it every evening. On Sankranti, when the sun enters Capricorn in the northern hemisphere, where the land of India is situated, the seeds are offered to the sun for his blessings, to ensure their fertility and a good harvest in the coming months. Seeds are closely connected with culture and women play a major role in their conservation. On the day of sowing, the women keep all the seeds meant for planting before the house deity and worship them. Before sowing begins, they worship the draft animals, the plough and other equipment used in agriculture. The acts of sowing and preservation are sacred, as are everything connected to nature and existence.

Seeds play an important role in Hindu rituals, ceremonies and festivals, celebrating the life cycle of birth, life, death and renewal. They are symbolic of fertility, eternity and sustenance in India. Seeds in general and *navadhanya* (nine sacred grains) in particular, symbolize the protection of biological and cultural diversity. Conserving seeds is conserving biodiversity, conserving knowledge of the seed and its utilization and conserving culture and sustainability. The sacred seeds are used in every rite and ritual: in the homa or havan, when a house is built,

Figure 7: Navadhanya—the nine sacred grains

Photograph by G. Balaji

when a baby is born, in the wedding ceremony and so on. No ritual is complete without the nine sacred seeds: wheat, Bengal gram, green gram, chickpea, white bean, black sesame, black gram, horse gram and rice (Figure 7). Thus, from birth to death, the produce of the earth is tended with care and respected at every ceremony. Her gift of new seeds every season is a gift of life itself.

'The Earth is my mother and I am her child,' says the Hymn to the Earth in the *Atharva Veda* (XII.1.12).

Every festival reminds us of the importance of nature in our lives. Many of them take place during certain phases of the moon or on specific asterisms and *nakshatra*s (lunar mansions). Each festival has a specific menu relating to the place, climate and the time of the year.

On 14 January, Hindus celebrate Makara Sankranti, when the sun transits to Capricorn in the northern hemisphere, marking the end of winter and the beginning of longer, warmer days. People all over India pray to the sun, the source of life and warmth. This festival indicates our ancient knowledge of astronomy. It is known as Lohri in the north, Bihu in Assam and Pongal or the celebration of harvest in the south. The day before Pongal is Bhogi, when old domestic items, such as mats, brooms and unwanted wooden furniture are burnt in a bonfire. The disposal of useless items is considered to be symbolic of the discarding of vices and attachment to material items. It is also an opportunity to throw away those items which can deteriorate, become a hideout for termites and insects, decompose and may even become a source of disease. Another festival—Holi—in the month of Phalgun (February–March) is a celebration of the spring harvest in north India and the fertility of the land. The *Holika dahan* held the previous night, when the old items in the house are burnt, is akin to the Bhogi fire held the day before Pongal. Akshaya Tritiya is yet another spring festival. While these are all solar festivities, Shivaratri in February–March, when people fast and meditate through the night, is a celebration of Shiva, the lunar deity.

Several festivals celebrate the role of Mother Earth and clay. In the hot summer months, the earthen tanks containing water are desilted and the discarded clay is used to make unbaked images of Ganesha which, after the festival, are put into a water source—sea, river or lake. Unfortunately, the original idea has been forgotten and today the idols are baked, made of plaster of Paris and painted with toxic colours, becoming a source of pollution. The Navaratri festival too celebrates clay, whether it is the beautiful clay images of Durga in Bengal and Bihar or the small clay *kolu* dolls of Tamil Nadu. Again, these simple figures of Durga have been supplanted by huge painted and elaborately decorated images of the goddess.

The discovery of fire gave human beings an advantage over animals, particularly large predators. Many festivals, such as Deepavali and Kartigai Deepam, celebrate agni and light. On the birthday of Rama and Krishna, many *deepam*s or diyas are lit in each home.

Flowers are used for decoration, but some festivals, such as Bathukamma in Telangana and Onam in Kerala, celebrate the flowers themselves. In Telangana, they appear in time for Navaratri and are made into elaborate *bathukammas*, which are later put into lakes, rivers and tanks, along with turmeric powder to clean the water source. Onam is the harvest festival of Kerala, which celebrates the annual return of Maharaja Bali to see his subjects. Hindus, Muslims and Christians make elaborate *pookalam*s on the floor with flowers, fruits and leaves to welcome him.

Tribal people in different states celebrate the diversity of nature and the importance of clay in elaborate festivals that are an integral part of their lives. Every Hindu festival celebrates its association with nature.

Once upon a time, India was a land of great biodiversity, with tropical rainforests and moist deciduous forests, savannah and grasslands. Tigers and lions, elephants and rhinoceros roamed the land. Today, forests are restricted to the Himalayan foothills and parts of north-east India, Bengal, the Western and Eastern Ghats, central and south India and the Andaman and Nicobar Islands. The forests are still the home of forest-dwelling tribes who have protected them as a matter of faith. Within these forests are sacred groves.

The forests are associated with a range of oral narratives and belief systems whereby the Mother Goddess—the Supreme Earth Mother—is generally the reigning deity of the forests, although there are some sacred groves with male gods too, such as Lord Vishnu, who is the tutelary deity of the groves in Assam.

The protection of forests is a part of Hindu dharma, the law of righteousness: 'Dharma exists for the welfare of all beings. Hence, that by which the welfare of all living beings is sustained, that for sure is dharma' (Mahabharata, 109.10).

The doctrine of karma tells us that every action will have its reaction: 'If one sows goodness, one will reap goodness; if one sows evil, one will reap evil.'

Hinduism has given great importance to forests. In Hindu texts, a fundamental sense of harmony with nature nurtured an ecological civilization. Forests are the primary source of life and inspiration, not a wilderness to be feared or conquered. The religious books of the Hindus were written by sages living in the forest who were motivated by a philosophy wherein the forest was a deliberate choice. They were the source of inspiration for the great rishis who composed some of the greatest verses of wisdom.

The human ability to merge with nature was the measure of cultural evolution. People drew intellectual, emotional and spiritual sustenance from the twin concepts of srishti or creation and prakriti or nature. 'So may the mountains, the waters, the plants, also heaven and earth, consentient with the Lord of the Forest (Vanaspati) preserve for us those riches' (*Rig Veda*, VII.34.23). Here the forest is regarded as the sovereign ruler of the earth.

Today, in the name of development, forests are being cut down with impunity. The waters are withdrawn in excess or polluted. The air is polluted with life-threatening chemicals. If the forests go, we go. If it disappears, our carbon sinks, our herbs and the food for so many species will disappear. We cannot live without water. We dig deeper into the earth in search of water, while our rivers have become streams of sewage and toxic chemicals. And, in the name of energy, we release toxic fumes and chemicals into the air. We set up animal farms with intensive breeding and culling of animals. Is this ethical? There is a total disregard and disrespect for our divinely procured natural resources.

There are also many different religious denominations living in the forests. It is, therefore, incumbent on all faiths to come together to end deforestation and protect the trees as a sacred duty.

The *Srimad Bhagavata Mahapurana* (2.2.41) states, 'Ether, air, fire, water, earth, planets, all creatures, directions, trees and plants, rivers and seas, they are all organs of God's body. Remembering this, a devotee respects all species.' Hindus must strive for ahimsa or non-violence in thought, word and deed, to minimize the harm we cause through our actions in our ordinary, day-to-day lives. As Hindus, we revere all life, human, non-human, plant and animal. The rivers and mountains are sacred. The planets and stars are celestial beings. When we embody this, we become servants of the Divine and all our actions, including those for protection of the world around us and all the beings therein, become acts of worship.

Aeons ago, the great Hindu sages realized that the harmony of the pancha-maha-bhuta—earth, air, fire, water and ether (prithvi, aapa, agni, vayu, akasha)—was essential to prevent climatic changes and disasters. If this harmony is realized again, we may be able to prevent greater calamities to this planet.

More than 5000 years ago, the scriptural foundation of Hinduism or Sanatana Dharma was laid down in the Vedas, the beginning of a continuous spiritual exploration. Today, Sanatana Dharma consists of many schools and teachers, each with a different perspective on humanity's relationship with the ultimate reality. Hinduism extols this pluralistic ethos through the Rigvedic verse *'ekam sad vipra bahudha vadanti'* (I.164) (Truth is one, the wise call it by many names).

The *Ishopanishad* (1.1) broadly characterizes the Hindu outlook: *'Ishavasyam idam sarvam'* (This entire universe belongs to the Lord). Therefore, take only what you need, what is set aside for you. Do not take anything else, for you know to whom it belongs.

Today, each of us must ask ourselves, 'How can I be of service? How can my service become an act of worship to honour and protect the earth?' For the last four decades, my colleagues and I have been documenting India's ecological traditions, protecting ancient forests and restoring sacred groves—fifty-three till date—with the help of local communities to whom they belong. Pilgrims sponsor tree plantations, creating new forests for Hindu temples. It is an effort to give back to the earth a small part of what we have taken from this planet.

Hinduism believes that the earth and all life forms—human, animal and plant—are a part of Divinity. Man evolved out of these life forms and is a part of the creative process, neither separate nor superior. We must change our lifestyle and habits to simplify our material desires, without taking more than our reasonable share of resources. Acting with an understanding of karma and the cycle of birth, death and rebirth are the central principles towards protecting the earth and the forests. 'Everything is divine,' says the *Bhagavad Gita*.

Through a combination of meaningful action, personal transformation and selfless service as an act of worship, we will be able to make the sort of inner and outer transitions that this planet requires. In doing so, we are acting in a deeply dharmic way, true to our Hindu ethos, philosophy and tradition.

At the beginning and end of the every Hindu ritual, this sloka is recited, uniting all of creation and praying for peace:

Om sham no Mitrah. Sham Varunah.
Sham no bhavatu Aryama.

Sham no Indro Brihaspatih.
Sham no Vishnur-urukramah

Namo Brahmane. Namaste Vaayo.
Tvameva pratyaksham Brahmaasi.
Tvaameva pratyaksham. Brahma vadishyaami.
Ritam vadishyaami. Satyam vadishyaami.

Tan-maamavatu. Tad-vaktaaram-avatu.
Avatu maam. Avatu vaktaaram.
Om shantih shantih shantih.
(*Taittiriya Upanishad, Anuvakah* 1)

Om

May the Guardian of the Cosmic Order be propitious
    towards us,
May the Waters be propitious towards us,
May the Sun be propitious towards us,
May the Rains and the Divine Teacher be propitious
    towards us,
May the long-striding Sun be Propitious towards us,

Salutations to the Supreme Creator
Salutations to the Breath of the Supreme Being,
You indeed are the Visible Creator,
I proclaim, You Indeed are the Visible Creator,
I speak of the Divine Truth, I speak of the Absolute Truth,

May That One protect me; May That One protect the
    Preceptor,
Protect me; Protect the Preceptor,
Om peace, peace, peace.

The *Atharva Veda* (XI.1.16) says that it is up to us, the progeny
of Mother Earth, to live in peace and harmony with all others:

'O Mother Earth! You are the world for us and we are your children; let us speak in one accord, let us come together so that we live in peace and harmony, and let us be cordial and gracious in our relationship with other beings.'

The Divine is all and all life is to be treated with reverence and respect. The family of Mother Earth—vasudhaiva kutumbakam—must promote 'sarva bhuta hita' (Yajur Veda, XII.32)—the welfare of all beings—people, animals and trees.

If forests and trees, fresh water and clean air disappear, so will all life on earth.

Acknowledgements

In 1992, I conceived the idea of conserving and restoring sacred groves in my home state of Tamil Nadu. That is when I started reading about and understanding how Hinduism is inexorably connected to nature, as is evident in ancient texts, faith practices and lifestyles. From there I went on to establishing a website ecoheritage.cpreec.org for C.P.R. Environmental Education Centre. In 2002, the Ministry of Environment and Forests and forests invited us to set up a website on 'Conservation of Ecological Heritage and Sacred Sites of India' as a part of the ministry's ENVIS (Environmental Information System) scheme, which led to the setting up of cpreecenvis. nic.in, and even won the Best ENVIS Centre Award in 2006. Subsequently, we conducted seminars in many Indian states and published the *Ecological Traditions of India* series.

Over the past few years, I have been attending several interfaith conferences all over the world where I have been speaking on Hinduism and nature. As a result, I have collected a wealth of material on the importance of nature and natural resources in Hinduism, which I have used in this book.

Most of the descriptions of groves, gardens, rivers, lakes and mountains come from my visits to sacred sites and temples, from Kailas in the north to Tirumala and Tiruvannamalai in the south, from the orans and vavs in the west to the raging Brahmaputra in the east and the great temples of Angkor Wat and Borobudur in South East Asia. I have also been able to obtain information from local priests and monks who were willing to share their knowledge and local folklore with me. The information about plants and animals come from my books *Sacred Animals of India* and *Sacred Plants of India*, respectively. I do not like to quote from websites. However, the paucity of information about many natural sites, especially mountains, forced me to go online. Most of what I have used here is taken from government, tourism and pilgrimage websites.

This book is a consequence of my two books—*Sacred Animals of India* (2010) and *Sacred Plants of India* (2014)— both published by Penguin India. Both received excellent reviews and, I believe, have been as popular as religious and environmental non-fiction can be. This opened my eyes to the wealth of reverence for nature in Indian religion and culture.

This book would not have been possible without the assistance of several people. H. Manikandan, my personal secretary, has been maintaining and finding files and papers, many misplaced by me from time to time, and protecting the entire project. My two librarians—Tamil Malar Kanthan of C.P.R. Institute of Indological Research and S.P. Vijayakumari of C.P.R. Environmental Education Centre—were invaluable in finding books, articles, journals and whatever else I required. M. Amirthalingam, my former student and research officer, C.P.R. Environmental Education Centre, and my co-author of *Sacred Plants of India*, gave me material and quotations from Tamil literature as well as botanical information and photographs. G. Balaji, assistant professor, C.P.R. Institute of Indological Research, collected the illustrations by Raja

Ravi Varma and others from the C.P. Ramaswami Aiyar Foundation's collections of art. V. Mohan, head, department of classical languages, CPR Institute of Indological Research, checked the Sanskrit translations in the text. B. Tirumala checked the references taken from our ENVIS website. R. Sathyanarayanan and M. Vaithiyanathan, our computer operators, C.P.R. Environmental Education Centre, formatted the references. The artists at the C.P. Ramaswami Aiyar Foundation—S. Prema and Y. Venkatesh—did the line drawings. And my husband, S. Chinny Krishna, my first editor and critic, did the proof correction and editing before I sent the manuscript to Penguin.

Finally, and most importantly, I initially discussed the project and signed the contract with Udayan Mitra, formerly with Penguin India, while Richa Burman and Aditi Muraleedharan, my editors at Penguin, had the arduous task of reading every word and line and making suggestions.

Thank you all very much.

Notes

Chapter 1: Introduction

1. M. Vannucci, *Ecological Readings in the Veda* (New Delhi: D.K. Printworld, 1994), p. 75.
2. Nanditha Krishna, *The Art and Iconography of Vishnu Narayana* (Bombay: D.B. Taraporevala Sons & Co., 1980), p. 24.
3. D.D. Kosambi, *An Introduction to the Study of Indian History* (Bombay: Popular Book Depot, 1956), pp. 70–72.
4. B.B. Lal, *The Sarasvatī Flows On: The Continuity of Indian Culture* (New Delhi: Aryan Books International, 2002), p. 75; Michel Danino, *The Lost River* (New Delhi: Penguin Books India, 2010), Chapter 11.
5. Ibid., pp. 67–68.
6. Stefano De Santis, *Nature and Man* (Varanasi: Sociecos & Dilip Kumar Publishers, 1995), p. 179.
7. Monier Monier-Williams, *Monier William's Sanskrit-English Dictionary*, 2nd ed. (Oxford: Oxford University Press, 1899), p. 654.
8. Vannucci, *Ecological Readings,* pp. 107–108, 112, 121.
9. O.P. Dwivedi, 'Dharmic Ecology', in *Hinduism and Ecology,* ed. C.K. Chapple and M.E. Tucker (New Delhi: Oxford University Press, 2000), p. 8.

10. Karan Singh, *Essays on Hinduism* (New Delhi: Ratna Sagar, 1995), p. 140.

11. *Ecological Traditions of Tamil Nadu,* ed. Nanditha Krishna (Chennai: C.P.R. Publications, 2005).

12. Danino, *The Lost River*, pp. 252–53.

13. Vannucci, *Ecological Readings,* pp. 87–90.

14. Karan Singh, *Essays.*

15. Ibid.

16. B. Prakash et al., 'Holocene tectonic movements and stress field in the western Gangetic plains', *Current Science* 79 (2000): 438–49.

17. Bridget Allchin and Raymond Allchin, *The Rise of Civilization in India and Pakistan* (Cambridge: Cambridge University Press, 1982), quoted in David L. Gosling, *Religion and Ecology in India and Southeast Asia* (London: Routledge, 2001).

18. 'India: Climate Change Impacts', worldbank.org. http://bit.ly/KOSc0Q.

19. S. Mohamed Imranullah, 'Religious beliefs protect nature: HC', *The Hindu,* 27 August 2017, Chennai edition.

Chapter 2: Groves and Gardens

1. C. Pathak, H. Mandalia and Y. Rupala, 'Bio-cultural Importance of Indian Traditional Plants and Animals for Environment Protection', *Review of Research* 1 (2012): 1–4.

2. R. Renugadevi, 'Environmental ethics in the Hindu Vedas and Puranas in India', *African Journal of History and Culture* (*AJHC*) 4 (2012): 1–3, http://www.academicjournals.org/ AJHC.

3. M.S. Umesh Babu and S. Nautiyal, 'Conservation and Management of Forest Resources in India: Ancient and Current Perspectives', *Natural Resources* 6 (2015): 256–72.

4. N.N. Sircar and R. Sarkar, *Vrksayurveda of Parasara* (A Treatise of Plant Science) (New Delhi: Sri Satguru Publications, 1996), pp. 13–18.

5. M. Amirthalingam and P. Sudhakar, *Plant and Animal Diversity in Valmiki's Ramayana* (Chennai: C.P.R. Publications, 2013), p. 13.

6. Ibid., pp. 13–16.

7. Nanditha Krishna, *The Book of Demons* (New Delhi: Penguin Books India, 2007).

8. S.N. Vyas, *India in the Ramayana Age* (New Delhi: Atmaram, 1967).

9. Ranchor Prime, *Hinduism and Ecology* (New Delhi: Motilal Banarsidass, 1994), pp. 17–18.

10. Romila Thapar, 'Perceiving the Forest: Early India', *Studies in History* (2001): pp.1–16.

11. B.S. Somashekar, 'Treasure House in Trouble', *Amruth* 2 (1998): pp. 3–7.

12. S. Rath, 'Kautilya's Attitude towards Fauna in the Arthasastra', in *Kautilya's Arthashastra and Social Welfare*, ed. V.N. Jha (New Delhi: Sahitya Akademi, 2006), pp. 279–80.

13. P. Sensarma, *Ethnobiological Information in Kautiliya Arthashastra* (Calcutta: Naya Prakash, 1998), pp. 10–27.

14. H. Skolmowski, 'Sacred groves in history', *Himalaya Man and Nature* XV (1991): 5.

15. M. Gadgil and V.D. Vartak, 'Sacred Groves of India: A Plea for Continued Conservation', *Journal of the Bombay Natural History Society* 73 (1975): 623–47.

16. S.A. Bhagwat and C. Rutte, 'Sacred groves: Potential for biodiversity management', *Frontiers of Ecology and the Environment* 4 (2006): 519–24.

17. Nanditha Krishna, 'Sacred Groves—An Indian Heritage', in *Sacred Groves of India—A Compendium*, ed. Nanditha Krishna and M. Amirthalingam (Chennai: C.P.R. Publications, 2015), pp. 71–79.

18. D.D. Kosambi, *Myth and Reality: Studies in the Formation of Indian Culture* (Bombay: Popular Prakashan, 1962).

19. Please see the C.P.R. Environmental Education Centre website, www.cpreecenvis.nic.in, for a state-wise list of sacred groves.

20. K.C. Malhotra et al., *Cultural and Ecological Dimensions of Sacred Groves in India* (New Delhi: Indian National Science Academy and Bhopal: Indira Gandhi Rashtriya Manav Sangrahalaya, 2001).

21. Chota Nagpur is a plateau in eastern India which covers most of Jharkhand state as well as adjacent parts of Odisha, West Bengal, Bihar and Chhattisgarh.

22. N. Jain, 'Community conservation in the Sikkim Himalaya', in *Community Conserved Areas in India—A Directory*, ed. Neema Pathak (Pune/Delhi: Kalpavriksh, 2009), pp. 629–40.

23. P. Medhi and S.K. Borthakur, 'Sacred groves of the Dimasas of North Cachar Hills', in *Sacred Groves of India—A Compendium*, ed. Nanditha Krishna and M. Amrithalingam (Chennai: C.P.R. Publications, 2013), pp. 71–79.

24. Sib Charan Roy J., *Ka Jingiapyni Ka Kmie Bad Ki Khun* (Shillong: Ri Khasi Press, rpt., 1993), p. 3.

25. S.K. Barik, B.K. Tiwari and R.S. Tripathi, *Sacred Groves of Meghalaya—A Scientific and Conservation Perspectives* (Shillong: NAEB, NEHU, 1998).

26. 'Khasi Hills—The Land of Abundant Waters', Mesmerizing Meghalaya: The Official Website of Meghalaya Tourism, megtourism.gov.in/dest-khasi.html.

27. R. Rai. 'Sacred Groves in Tribal Pockets of Madhya Pradesh and Plants Conserved by Ethnic Communities in Conservation of Biodiversity', in *Ecological Traditions of Madhya Pradesh and Chhattisgarh* IX, ed. Nanditha Krishna (Chennai: C.P.R. Environmental Education Centre and www.cpreecenvis.nic.in, 2014), pp. 80–96.

28. S. Patnaik and A. Pandey, 'A study of indigenous community based forest management system: Sarna (sacred groves)', in *Conserving the Sacred* for *Biodiversity Management*, ed. P.S. Ramakrishnan et al. (New York: Oxford and IBH, 1998), pp. 315–22,.

29. Kosambi, *Myth and Reality*; M. Gadgil and V.D. Vartak, 'Sacred groves of Western Ghats of India', *Economic Botany* 30 (1976): 152–60; M. Gadgil and V.D. Vartak, 'Sacred groves in

Maharashtra—An Inventory', in *Glimpses of Indian Ethnobotany*, ed. S.K. Jain (New Delhi: Oxford and IBH, 1981), pp. 279–94.

30. G.D. Sontheimer, 'Hinduism: The Five Components and Their Interaction', in *Hinduism Reconsidered*, ed. G.D. Sontheimer and H. Kulke (New Delhi: Manohar, 1989).

31. R. Rajamani, 'Ecological Traditions of Andhra Pradesh', *Ecological Traditions of Andhra Pradesh* (Chennai: C.P.R. Publications, 2005), p. 11.

32. C.J. Sonowal and P. Praharaj, 'Tradition vs Transition: Acceptance of Health Care Systems among the Santhals of Orissa', *Ethno-Med* 1 (2007): 135–46.

33. P. Dayanandhan, 'Origin and meaning of the Tinai concept in Sangam Tamil Literature', in *Indological Essays, Commemorative Volume II for Gift Sironmoney*, ed. Michael Lockwood (Chennai: Dept. of Statistics, Madras Christian College, 1992), pp. 27–44.

34. G. Karunanithi Arunachalam et al., 'Ethno Medicines of Kolli Hills at Namakkal District in Tamilnadu and its significance in Indian Systems of Medicine', *Journal of Pharmaceutical Science & Research* 1 (2009): 1–15.

35. Nanditha Krishna, 'The terracotta tradition of the sacred groves', in *The Ecological Traditions of Tamil Nadu*, ed. N. Krishna and J. Prabhakaran (Chennai: C.P.R. Environmental Education Centre, 1997), pp. 62–65.

36. 'South Indian Inscriptions: Pallava Inscriptions Nos 101 to 125', whatisindia.com, http://bit.ly/2yinaZd.

37. 'South Indian Inscriptions: Inscription Collected during the Year 1908–09', whatisindia.com, http://bit.ly/2yfmkyE.

38. 'South Indian Inscriptions Volume XVII: Inscriptions Collected during 1903–04', whatisindia.com, http://bit.ly/2g5WKl5.

39. 'South Indian Inscriptions: Inscriptions of the Ranganathasvamy Temple, Srirangam', whatisindia.com, http://bit.ly/2kKBJ4H.

40. 'South Indian Inscriptions: Inscriptions Collected during the Year 1906', whatisindia.com, http://bit.ly/2zi3uF3.

41. M. Amirthalingam, *Sacred Groves of Tamil Nadu—A Survey* (Chennai: C.P.R. Environmental Education Centre, 1998).

42. Nanditha Krishna, *The Arts and Crafts of Tamilnadu* (Ahmedabad: Mapin Publishing, 1992), p. 90.

43. S. Inglis, *A Village Art of South India* (Madurai: Madhurai Kamaraj University, 1980).

44. A. Godbole et al., 'Role of sacred groves in biodiversity conservation with local people's participation: A case study from Ratnagiri district, Maharashtra', in *Conserving the Sacred for Biodiversity Management,* ed. P.S. Ramakrishnan, K.G. Saxena and U.M. Chandrashekara (New Delhi: Oxford & IBH, 1998), pp. 233–46.

45. J.J. Roy Burman, 'A comparison of sacred groves among the Mahadeo Kolis and Kunbis of Maharashtra', *Indian Anthropologist* 26 (1996): 37–46.

46. Gadgil and Vartak, 'Sacred Groves in Maharashtra', pp. 279–94.

47. M. Gadgil, 'Conserving biodiversity as if people matter: A case study from India', *Ambio* 21 (1992): 266–70.

48. V. Sarojini Menon, *Sacred Groves: The Natural Resources of Kerala* (Trivandrum: WWF, Kerala Office, 1997).

49. Gadgil and Vartak, 'Sacred groves in Maharashtra'.

50. S.A. Bhagwat and C. Rutte, 'Sacred groves: Potential for biodiversity management', *Frontiers of Ecology and the Environment* 4 (2006): 519–24.

51. R.S. Tripathi, B.K. Tiwari and S.K. Barik, *Sacred Groves of Meghalaya: Status and Strategy for Their Conservation* (Shillong: NAEB, NEHU, 1998), pp. 112–25; B.K. Tiwari, S.K. Barik and R.S. Tripathi, 'Biodiversity value, status, and strategies for conservation of sacred groves of Meghalaya, India', *Ecosystem Health* 4 (1998): 22–33.

52. V.D. Vartak, 'Sacred Groves of Tribals for In-Situ Conservation of Biodiversity', in *Ethnobiology in Human Welfare,* ed. S.K. Jain (New Delhi: Deep Publications, 1996), pp. 300–02.

53. S. Sukumaran and A.D.S. Raj, 'Sacred groves as a symbol of sustainable environment—A case study', in *Sustainable*

Environment, ed. N. Sukumaran (Thirunelveli: SPCES, MS University, Alwarkurichy, 1999), pp. 67–74.

54. B. Maheswaran, P. Dayanandan and D. Narasimhan, 'Miniature Sacred Grove near Vedanthangal Bird Sanctuary', in *Abstracts of 2nd Congress on Traditional Science and Technology of India*, Madras, Bio 3 (1995): 26–31.

55. M. Gadgil and V.D. Vartak, 'Sacred groves of India: A Plea for Continued Conservation', *Journal of Bombay Natural History Society* 72 (1975): 314–20; M.P. Ramanujam, 'Conservation of environment and human rights; sacred groves in cultural connections to biodiversity PRP', *Journal of Human Rights* 4 (2000): 34–38.

56. M.A. Kalam, *Sacred Groves in Kodagu District of Karnataka (South India): A Socio Historical Study* (Pondicherry: Institut Français de Pondichéry, 1996).

57. M.D.S. Chandran and J.D. Hughes, 'The Sacred Groves of South India: Ecology, Traditional Communities and Religious Change', *Social Compass* 44 (1997): 413–27.

58. Much of this section is taken from www.cpreecenvis.in.

59. M. Amirthalingam, 'Conservation as a Tamil Ethic', *ECONEWS* XII (2006): 22–25.

60. For more information on the species protected in each state, visit the C.P.R. Environmental Education Centre website, www.cpreecenvis.nic.in.

Chapter 3: Divine Waters

1. A.B. Keith, *The Religion and Philosophy of the Veda and Upanisad* (New Delhi: Motilal Banarsidass, 1970), pp. 141–42.

2. Sanjeev Sanyal, *Land of the Seven Rivers* (Gurgaon: Penguin Books India, 2012).

3. Michel Danino, *The Lost River* (New Delhi: Penguin Books India, 2010).

4. 'Beas', sikhiwiki.org. http://www.sikhiwiki.org/index.php/Beas.

5.   Dr Ved Kumari, *Nilamata Purana* (Srinagar: Jammu and Kashmir
     Academy of Art, Culture and Languages, 1968), verses 247–61.
6.   Frederic Eden Pargiter, *Ancient Indian Historical Traditions* (London:
     Oxford University Press, H. Milford, 1852–1927), pp. 172–82.
7.   S. Vaidyanathan and Shayoni Mitra, *Rivers of India* (New Delhi:
     Niyogi Books, 2011), p. 60.
8.   Diana L. Eck, 'India's "Tírthas": "Crossings" in Sacred
     Geography', *History of Religions* 20 (1981): 323–44.
9.   Diana L. Eck, *Banaras: City of Light* (New York: Columbia
     University Press, 1998), pp. 145–46.
10. Ibid., p. 217.
11. Vandana Shiva, *Earth Democracy: Justice, Sustainability and Peace*.
     G-Reference, Information and Interdisciplinary Subjects Series
     (London: Zed Books Ltd, 2006), pp. 172–73.
12. Allan Dahlaquist, 'Megasthenes and Indian Religion', *History
     and Culture,* Vol. 11 (New Delhi: Motilal Banarsidass, 1996),
     p. 386.
13. Govind Singh, Mihir Deb and Chirashree Ghosh, 'Urban
     Metabolism of River Yamuna in the National Capital Territory
     of Delhi, India', *International Journal of Advanced Research* 4 (2016):
     1240–48.
14. 'A First In India: Uttarakhand HC Declares Ganga, Yamuna
     Rivers as Living Legal Entities', LiveLaw.in, http://bit.
     ly/2nYLFV9.
15. Pushkaram is celebrated at shrines along the banks of twelve
     major sacred rivers in India.
16. R. Biswas, *Brahmaputra and the Assam Valley* (New Delhi:
     Niyogi Books, 2013).
17. 'River Brahmaputra', C.P.R. Environmental Education Centre
     website, http://www.cpreecenvis.nic.in/Database/River_
     Brahmaputra_891.aspx.
18. Ibid.
19. Vaidyanathan and Shayoni. *Rivers,* pp. 138–40.
20. Ibid., pp. 98–100.

21. Padma Seshadri and Padma Malini Sundararaghavan, *It Happened along the Kaveri* (New Delhi: Niyogi Books, 2012), pp. 18, 22.

22. *The Hindu*, 'Vatarangam, seat of Hari and Haran', http://bit.ly/2yXKScj.

23. P.V. Kane, *History of Dharmasastra: Ancient and Medieval Religious and Civil Law in India* (Pune: Bhandarkar Oriental Research Institute, 1953), pp. 554–55.

24. Savitri V. Kumar, *The Puranic Lore of Holy Water Places* (New Delhi: Munshiram Manoharlal, 1983), pp. 9–10.

25. Nitya Jacob, *Jalyatra* (New Delhi: Penguins Books India, 2008), p. 5.

26. 'Punjab', C.P.R. Environmental Education Centre website, http://www.cpreecenvis.nic.in/Database/Punjab_941.aspx.

27. Ibid.

28. 'Arunachal Pradesh', C.P.R. Environmental Education Centre website, http://www.cpreecenvis.nic.in/Database/Arunachal_1764.aspx.

29. P. Ganguli et al., 'Prospects of Ecotourism in Temple Tanks and Floodplain Lakes of Upper Assam', *Proceedings of TAAL 2007: The 12th World Lake Conference* (2008): 1329–32.

30. Sairem Nilabir, *Laiyingthou Sanamahi Amasung Sanamahi Laining Hinggat Eehou* (Imphal: Thingbaijam Chanu Sairem Ongbi Ibemhal, 2002), p. 102.

31. H.H. Mohrmen, 'Heritage sites in Jaintia hills', *Shillong Times*, 30 July 2012.

32. C. Acharya and A.S. Gyatso Dokham, 'Sikkim: The Hidden holy land and its sacred lakes', in *Bulletin of Tibetology,* ed. Acharya Samten Gyatso, Rigzin Nyodup and Thupten Tenzing (1998): 10–15.

33. Draft Management Plan for Tsomgo Lake, Department of Forest, Environment and Wildlife Management, Government of Sikkim, August 2008.

34. State of the Environment Report, Ministry of Environment and Forests, Sikkim, 2007, p. 56–57.

35. W.W. Hunter, *The Statistical Account of Bengal* (London: Trubner & Co., 1875).

36. G.K. Bera, 'Temples, Fairs and Festivals of Tripura', *TUI, Journal of Tribal Life and Culture* 16 (2013): 108.

37. 'District Profile: Dhalai, Tripura', 2015, p. 4, http://bit.ly/2z4uirx.

38. D. Deb, 'Sacred Ecosystem of West Bengal', in *Status of Environment in West Bengal: A Citizens' Report*, ed. A.K. Ghosh (Kolkata: ENDEV—Society for Environment and Development, 2008), p. 122.

39. M. Praharaj, 'Historic Conservation and Sustainability: A Case of Bindusagar Lake, Old Bhubaneswar', *Elixir Sustain. Arc.* 51A (2012): 1090–93.

40. S.K. Rath, 'Narendra Tank in Legend and History', *Orissa Review*, June issue (2004): 13–15.

41. Nanditha Krishna, *Paintings of the Varadaraja Perumal Temple, Kanchipuram* (Chennai: C.P.R. Publications, 2014), pp. 13–15, 86.

42. Ibid.

43. Nanditha Krishna, *Balaji-Venkateshwara, Lord of Tirumala-Tirupati—An Introduction* (Mumbai: Vakils, Feffer and Simons, 2000), p. 21.

44. Savitri V. Kumar, *The Puranic Lore of Holy Water Places* (New Delhi: Munshiram Manoharlal, 1983), p. 231.

45. 'Sacred Waterbodies of India', ecoheritage.cpreec.org, Ecoheritage.cpreec.org/Sacred Waterbodies of India.

46. M. Amirthalingam, 'Chennai's Changing Landscape', *ECONEWS* 15 (2009): 23–26.

Chapter 4: Plants as Protectors

1. For more information on this subject, please see Nandita Krishna and M. Amrithalingam, *Sacred Plants of India* (Gurgaon: Penguin Books India, 2014), which is the source for much of this chapter.

2. Surapala, *Vrukshayurveda,* trans. Nalini Sadhale, *Agri-History Bulletin* No. 1 (Secunderabad: Asian Agri-History Foundation, 1996).

3. O.P. Dwivedi, *World Religions and the Environment* (New Delhi: Gitanjali Publishing House, 1989), p. 175.

4. P.S. Ramakrishnan, 'Sacred Ecological History of India: Ancient to the Contemporary', in *The Environment and Indian History,* ed. Nanditha Krishna (Chennai: CPR Publications, 2016), p. 71.

5. Kamala Vasudevan, 'Tree as sacred symbol', *Chitralakshana,* www.chitralakshana.com/trees.html.

6. M. Vannucci, *Ecological Readings in the Veda* (New Delhi: D.K. Printworld, 1994), pp. 105–06.

7. 'Aswatha Vruksha Stotram', trans. P.R. Ramachander, hindupedia.com, http://bit.ly/2hhMYxh.

8. Z.A. Ragozin, *Vedic India as Embodied Principally in the Rig Veda* (New Delhi: Munshiram Manoharlal, 1961), p. 159.

9. The khejri is the sacred plant of the Bishnois who live in the deserts of Rajasthan. They were instructed to worship this tree, essential for desert ecology and their own survival, by their religious leader Shree Guru Jambeshwar Bhagwan, also known as Jamboji (1451–1536 CE), in the sixteenth century.

10. Sumati Krishnan, 'Tulasi Jagajanani', *The Hindu Friday Review,* 30 June 2017.

11. M. Amirthalingam and P. Sudhakar, *Plant and Animal Diversity in Valmiki's Ramayana* (Chennai: C.P.R. Environmental Education Centre, 2013), p. 18.

12. Nanditha Krishna, *The Art and Iconography of Vishnu-Narayana* (Bombay: D.B. Taraporevala Sons & Co., 1980), p. 7.

13. A.K. Coomaraswamy, *Yaksas* (New Delhi: Munshiram Manoharlal, 1980). I.14–16; II.1–12.

14. Ibid.

15. Ibid., II.7.

16. A.L. Srivastava, *Life in Sanchi Sculpture* (New Delhi: Abhinav Publications, 1983), p. 135.

17. P.L. Kaler, 'Sacred Trees of Punjab', in *Ecological Traditions of India*, Vol. VIII, ed. Nanditha Krishna (Chennai: C.P.R. Environmental Education Centre), pp. 95–101; and www.cpreecenvis.nic.in.

18. M. Amirthalingam, 'Ecosystem Services Provided by Sacred Plants', *ENVIS Newsletter* XV: 2–4, www.ecoheritage.com.

Chapter 5: Children of Pashupati

1. For more information on this subject, please see: Nanditha Krishna, *Sacred Animals of India* (New Delhi: Penguin Books India, 2010), which is the source for much of this chapter.

2. C. Chapple, *Nonviolence to Animals, Earth and Self in Asian Traditions* (New York: State University of New York Press, 1993), Chapter 1.

3. P. Swami, *Encyclopaedic Dictionary of Upanisads* (New Delhi: Sarup & Sons, 2000), Vol. 3, pp. 630–31.

4. C. Chapple, 'Ecological Nonviolence and the Hindu Tradition', *Perspectives on Nonviolence* (New York: Springer, 1990), pp. 168–77.

5. Maneka Sanjay Gandhi, 'Thirukkural on the virtues of vegetarianism', *Mathrubhumi,* 6 March 2017.

6. Cf. Nanditha Krishna, *Sacred Animals.*

7. Nanditha Krishna, *The Book of Demons* (New Delhi: Penguin Books India, 2007).

8. Harriet Ritvo, 'Beasts in the Jungle (or wherever)', *Dedalus: The Journal of the American Academy of Arts and Sciences*, Spring 137 (2008): pp. 22–30.

9. Nanditha Krishna, *Sacred Animals.*

10. Ibid., p. 78.

11. Ibid., p. 5.

12. Manoj R. Borkar, 'Cultural Relics of Eco-centrism in Goa—An Eco-anthropological Analysis', in *Ecological Traditions of Goa*, ed. Nanditha Krishna (Chennai: CPR Environmental Education Centre, 2010).

Chapter 6: Abode of the Gods

1. Most of the material on individual mountains is from www. cpreecenvis.nic.in, which we collected from all the states of India to upload on the website. However, only the important mountains have been included in this book.

2. Kalidasa, *Kumarasambhavam*, trans. Rajendra Tandon (New Delhi: Rupa & Co, 2008).

3. 'Sacred Mountains', C.P.R. Environmental Education Centre website, www.cpreecenvis.nic.in/sacred mountains.

4. John Snelling, *The Sacred Mountain* (New Delhi: Motilal Banarsidass, 2006), pp. 22–23.

5. Charles Allen, *A Mountain in Tibet* (London: Futura Publications, 1991), pp. 21–22.

6. Arthur Avalon and Sir John Woodroffe, *The Tantra of the Great Liberation* (Whitefish, Montana: Kessinger Publishing, 2004), Ibid.

7. Chandra Rajan, *The Complete Works of Kalidasa*, Vol. 1 (New Delhi: Sahitya Akademi, 1997), p. 307.

8. Kalidasa, 'Purva Megha', *Meghasandesa* (Madras: Balamanorama Press, 1957), 58.

9. 'Hari Parbat', C.P.R. Environmental Education Centre website, www.cpreecenvis.nic.in/Database/HariParbat.

10. Ritu Singh et al., *Manimahesh Sacred Landscape—A Monograph* (New Delhi: INTACH, 2016).

11. Kuldeep Chauhan, 'Raiding the Himalayas', *Tribune*, Chandigarh, 20 September 2006.

12. 'Haryana: Dhosi Hills', C.P.R. Environmental Education Centre website, www.cpreecenvis.nic.in/Database/ DhosiHill.

13. 'Bandarpunch (6316m)', whitemagicadventure.com, http://bit. ly/2gfxT23.

14. 'Joshimath', trekearth.com, http://bit.ly/2yemhmW.

15. 'Dunagiri', indiamapped.com, http://bit.ly/2kLvqxD.

16. 'Himavat', freeBSD.nfo.sk, http://bit.ly/2z4f9Gu.

17. 'Bagheshwari Temple', BhaaratDarshan.com, http://bit.ly/2z3PbmE.
18. 'Navagraha temple in Guwahati, Assam', musetheplace.com, http://bit.ly/2gAtUKs.
19. 'Temples of Assam', online.assam.gov.in, http://bit.ly/2kJVsl8.
20. 'History of Various Than Institutions in Kamrup District', shodhganga.inflibnet.ac.in, http://bit.ly/2yIgQ0j.
21. 'Temples of Assam', online.assam.gov.in, http://bit.ly/2kJVsl8.
22. 'Basistha Temple, Guwahati', templepurohit.com, http://bit.ly/2yJ4hSf.
23. 'Place of Interest', The Official Website of Manipur State, http://bit.ly/2yIokjP.
24. 'Jaintia Hills—Land of Myths and Legends', The Official Website of Meghalaya Tourism, http://megtourism.gov.in/dest-jaintia.html.
25. 'Mount Kangchenjunga', C.P.R. Environmental Education Centre website, http://bit.ly/2gBPgHa.
26. 'Guru Padmasambhava', The Official Website of Guru Padmasambhava Sikkim, http://samdruptse.nic.in.
27. 'Tendon Hill', holidayiq.com, http://bit.ly/2zjM6zA.
28. 'Tripura: Unakoti Hill', C.P.R. Environmental Education Centre website, www.cpreecenvis.nic.in/Database/Tripura
29. 'Kalimpong Attractions & Activities', darjeeling-tourism.com, http://bit.ly/2i6poXv.
30. 'Sita Kund', The Official Website of Purulia, http://bit.ly/2gC6c0t.
31. 'Mama-Bhagne Pahar', panoramio.com, http://bit.ly/2yIimQ3.
32. Juhi Chaudhary, 'Interview: Niyamgiri Activist Gets World's Biggest Green Prize', *Wire*, 25 April 2017, http://bit.ly/2hEuExU.
33. 'The History of Mandar Hill and its Religious Significance', explorbihar.in, http://bit.ly/2gghMkC.
34. 'Kauleshwari Hill', chatra.nic.in, http://bit.ly/2yl4XMA.

35. 'Introduction', The Official Website of Parasnath: Giridih Tourism, http://bit.ly/2i9CxPI.

36. 'Chitrakoot: Mythology Revisited', *Outlook Traveller*, 15 May 2017, http://bit.ly/2xDepqM.

37. 'The Temple: Ambaji', The Official Website of Shri Arasuri Ambaji Mata Devasthan Trust (SAAMDT), http://bit.ly/2g7EgAL.

38. *Skanda Purana*, 4.1–3; also Savitri Kumar, *The Pauranic Lore of Holy Water-places* (Delhi: Munshiram Manoharlal, 1983), p. 183.

39. Cynthia Ann Humes, 'Vindhyavasini: Local Goddess yet Great Goddess', in *Devi: Goddesses of India*, ed. John Stratton Hawley and Donna M. Wulff (Delhi: Motilal Banarsidass, 1998), p. 49.

40. Atul Dev, 'The Unending Fallout of Unilever's Thermometer Factory in Kodaikanal', *Caravan—A Journal of Politics and Culture*, 20 August 2015.

41. P.V. Jagadisa Aiyar, *South Indian Shrines* (New Delhi: Asian Educational Services, 1982), pp. 191–203.

Select Bibliography

Primary Sources

Arthashastra. trans. R. Shamasastry. Mysore: Wesleyan Mission Press, 1929.

Atharva Veda. Hoshiarpur: Vishveshvaranand Indological Series, 1960–64.

Bhagavad Gita. Mumbai: The Bhaktivedanta Book Trust, 1986.

Brihadaranyaka Upanishad. Poona: Vaidika Samsodhana Mandala, 1958.

Chandogya Upanishad. Poona: Vaidika Samsodhana Mandala, 1958.

Charaka Samhita. Varanasi: Chowkhamba Sanskrit Series, 2003–05.

Durga Saptashati. Gorakhpur: Gita Press, 2013.

Ishopanishad. Poona: Vaidika Samsodhana Mandala, 1958.

Kumarasambhavam by Kalidasa. trans. Rajendra Tandon. New Delhi: Rupa & Co., 2008.

Mahabharata. Calcutta: Asiatic Society of Bengal, 1837; trans. M.N. Dutt. New Delhi: Piramal Publications, 1988.

Manu Samhita. Calcutta: Education Press, 1830.

Mundakopanishad. Poona: Vaidika Samsodhana Mandala, 1958.

Nirukta. ed. and trans. Lakshman Sarup. New Delhi: Motilal Banarsidass, 1967.

Purananuru, Ettutogai. Chennai: Pari Nilayam, 2009.

Ramayana. trans. R.T.H. Griffith. Varanasi: Chowkhamba Sanskrit Studies, 1963; Chandrasekhara Aiyar, N. Chennai: C.P.R. Publications, 2013.

Rig Veda. Hoshiarpur: Vishveshvaranand Indological Series, 1963–66; trans. R.T.H. Griffith. Benares: E.J. Lazarus and Co., 1896.

Shatapatha Brahmana. trans. J. Eggeling. Sacred Books of the East. Delhi: Motilal Banarsidass, 1963–66.

Shvetashvatara Upanishad. Poona: Vaidika Samsodhana Mandala, 1958.

Silappadikaram. Madras: Pari Nilayam, 1965.

Savitri V. Kumar. *The Pauranic Lore of Holy Water-places with Special Reference to Skanda Purana.* New Delhi: Munshiram Manoharlal, 1983.

Srimad Bhagavata Mahapurana. Gorakhpur: Gita Press, 1971.

Taittiriya Upanishad. Bombay: Central Chinmaya Mission Trust, 1995.

Agni Purana. Poona: Anandashrama Sanskrit Series, 1900.

Vishnu Purana. Commentary by Shridhara Svamin. Bombay: Shri Venkateshwara Steam Press, 1967.

Vrukshayurveda by Surapala. trans. Nalini Sadhale. *Agri-History Bulletin* 1. Secunderabad: Asian Agri-History Foundation, 1966.

Yajur Veda Samhita. Ajmer: Shrimad Paropakarini Sabha, 2007.

Secondary Sources

Acharya, C., and A.S. Gyatso Dokham. 'Sikkim: The Hidden holy land and its sacred lakes'. In *Bulletin of Tibetology*. Edited by Acharya Samten Gyatso, Rigzin Nyodup and Thupten Tenzing. Gangtok: Namgyal Institute of Tibetology, 1998.

Aiyar, P.V. Jagadisa. *South Indian Shrines.* New Delhi: Asian Educational Services, 1982.

Allchin, Bridget and Raymond Allchin. *The Rise of Civilization in India and Pakistan.* Cambridge: Cambridge University Press, 1982.

Allen, Charles. *A Mountain in Tibet.* London: Futura Publications, 1991.

Amirthalingam, M. *Sacred Groves of Tamil Nadu—A Survey*. Chennai: C.P.R. Environmental Education Centre, 1998.

Amirthalingam, M. 'Chennai's Changing Landscape'. *ECONEWS* XV (2009).

Amirthalingam, M. 'Ecosystem Services Provided by Sacred Plants'. *ECONEWS* XVIII (2013).

Amirthalingam, M., and P. Sudhakar. *Plant and Animal Diversity in Valmiki's Ramayana*. Chennai: C.P.R. Publications, 2013.

Annual Reports (AR), Archaeological Survey of India (ASI).

Barik, S.K., B.K. Tiwari and R.S. Tripathi. *Sacred Groves of Meghalaya—A Scientific and Conservation Perspective*. Shillong: Regional Centre, NAEB, NEHU, 2006.

Bera, G.K. 'Temples, Fairs and Festivals of Tripura', *TUI, Journal of Tribal life and Culture* 16 (2013).

Bhagwat, S.A. and C. Rutte. 'Sacred groves: Potential for biodiversity management'. *Frontiers of Ecology and the Environment* 4 (2006).

Biswas, R. *Brahmaputra and the Assam Valley*. New Delhi: Niyogi Books, 2013.

Borkar, Manoj R. 'Cultural Relics of Eco-centrism in Goa—An Eco-anthropological Analysis'. In *Ecological Traditions of Goa*. Edited by Nanditha Krishna. Chennai: C.P.R. Environmental Education Centre, 2010.

Burman, J.J. Roy. 'A comparison of sacred groves among the Mahadeo Kolis and Kunbis of Maharashtra'. *Indian Anthropologist* 26 (1996): 37–46.

Chandran, M.D.S. and J.D. Hughes. 'The Sacred Groves of South India: Ecology, Traditional Communities and Religious Change'. *Social Compass* 44 (1997): 413–27.

Chapple, C. 'Ecological Nonviolence and the Hindu Tradition'. In *Perspectives on Nonviolence*. Edited by V.K. Kool. New York: Springer, 1990.

Chapple, C. *Nonviolence to Animals, Earth and Self in Asian Traditions*. New York: State University of New York Press, 1993.

Dahlaquist, Allan. 'Megasthenes and Indian Religion'. *History and Culture*. Vol. 11. New Delhi: Motilal Banarsidass, 1996.

Dayanandhan, P. 'Origin and meaning of the Tinai concept in Sangam Tamil Literature'. In *Indological Essays, Commemorative Volume II for Gift Sironmoney*. Edited by Michael Lockwood. Chennai: Department of Statistics, Madras Christian College, 1992.

Deb, D. 'Sacred Ecosystem of West Bengal'. In *Status of Environment in West Bengal: A Citizens' Report*. Edited by A.K. Ghosh. Kolkata: ENDEV—Society for Environment and Development, 2008.

De Santis, Stefano. *Nature and Man*. Varanasi: Sociecos & Dilip Kumar, 1995.

Dwivedi, O.P. *World Religions and the Environment*. New Delhi: Gitanjali Publishing House, 1989.

Dwivedi, O.P. 'Dharmic Ecology'. In *Hinduism and Ecology*. Edited by C.K. Chapple and M.E. Tucker. New Delhi: Oxford University Press, 2000.

Eck, Diana L. 'India's Tírthas: Crossings in Sacred Geography'. *History of Religions* 20 (1981).

Eck, Diana L. *Banaras: City of Light*. New York: Columbia University Press, 1998.

Gadgil, M. 'Conserving biodiversity as if people matter: A case study from India'. *Ambio* 21 (1992).

Gadgil, M. and V.D. Vartak. 'Sacred Groves of India: A Plea for Continued Conservation'. *Journal of the Bombay Natural History Society* 73 (1975).

Gadgil, M. and V.D. Vartak. 'Sacred groves in Maharashtra—An inventory'. In *Glimpses of Indian Ethnobotany*. Edited by S.K. Jain. New Delhi: Oxford & IBH, 1981.

Godbole, A., A. Watve, S. Prabhu and J. Sarnaik. 'Role of sacred groves in biodiversity conservation with local people's participation: A case study from Ratnagiri district, Maharashtra'. In *Conserving the Sacred for Biodiversity Management*. Edited by P.S. Ramakrishnan, K.G. Saxena and U.M. Chandrashekara. New Delhi: Oxford & IBH, 1998.

Gosling, David L. *Religion and Ecology in India and Southeast Asia*. London: Routledge, 2001.

Humes, Cynthia Ann. 'Vindhyavasini: Local Goddess yet Great Goddess'. In *Devi: Goddesses of India*. Edited by John Stratton Hawley and Donna M. Wulff. Delhi: Motilal Banarsidass, 1998.

Inglis, S. *A Village Art of South India*. Madurai: Madurai Kamaraj University, 1980.

Jacob, Nitya. *Jalyatra*. New Delhi: Penguins Books India, 2008.

Jain, N. 'Community conservation in the Sikkim Himayalaya'. In *Community Conserved Areas in India—A Directory*. Edited by Neema Pathak. Pune/Delhi: Kalpavriksh, 2009.

Kalam, M. A. *Sacred Groves in Kodagu District of Karnataka (South India): A Socio Historical Study*. Pondicherry: Institut Français de Pondichéry, 1996.

Kaler, P.L. 'Sacred Trees of Punjab'. In *Ecological Traditions of Punjab*. Edited by Nanditha Krishna. Chennai: C.P.R. Environmental Education Centre, 2014.

Kane, P.V. *History of Dharmasastra: Ancient and Medieval Religious and Civil Law in India*. Pune: Bhandarkar Oriental Research Institute, 1953.

Karan Singh. *Essays on Hinduism*. New Delhi: Ratna Sagar, 2014.

Keith, A.B. *The Religion and Philosophy of the Veda and Upanisad*. New Delhi: Motilal Banarsidass, 1970.

Kosambi, D.D. *An Introduction to the Study of Indian History*. Bombay: Popular Book Depot, 1956.

Kosambi, D.D. *Myth and Reality: Studies in the Formation of Indian Culture*. Bombay: Popular Prakashan, 1962.

Krishna, Nanditha. *The Art and Iconography of Vishnu Narayana*. Bombay: D.B. Taraporevala Sons & Co., 1980.

Krishna, Nanditha. *The Arts and Crafts of Tamilnadu*. Ahmedabad: Mapin, 1992.

Krishna, Nanditha, ed. *The Ecological Traditions of Tamil Nadu*. Chennai: C.P.R. Environmental Education Centre, 1997.

Krishna, Nanditha. *Balaji-Venkateshwara, Lord of Tirumala-Tirupati—An Introduction*. Mumbai: Vakils, Feffer and Simons, 2000.

Krishna, Nanditha. *The Book of Demons*. New Delhi: Penguin Books India, 2007.

Krishna, Nanditha. *Sacred Animals of India*. New Delhi: Penguin Books India, 2010.

Krishna, Nanditha. *Paintings of the Varadaraja Perumal Temple, Kanchipuram*. Chennai: C.P.R. Publications, 2014.

Krishna, Nanditha, ed. *The Environment and Indian History*. Chennai: C.P.R. Publications, 2016.

Krishna, Nanditha and M. Amirthalingam. *Sacred Plants of India*. Gurgaon: Penguin Books India, 2014.

Krishna, Nanditha and M. Amirthalingam, ed. *Sacred Groves of India—A Compendium*. Chennai: C.P.R. Publications, 2015.

Kumar, Savitri V. *The Puranic Lore of Holy Water Places*. New Delhi: Munshiram Manoharlal, 1983.

Maheswaran, B., P. Dayanandan and D. Narasimhan. 'Miniature Sacred Grove near Vedanthangal Bird Sanctuary'. In *Abstracts of Traditional Science and Technology of India*, Madras, Bio 3 (1995).

Malhotra, K.C., Y. Gokhale, S. Chatterjee and S. Srivastava. *Cultural and Ecological Dimensions of Sacred Groves in India*. New Delhi/Bhopal: Indian National Science Academy/Indira Gandhi Rashtriya Manav Sangrahalaya, 2001.

Menon, V. Sarojini. *Sacred Groves: The Natural Resources of Kerala*. Trivandrum: WWF, Kerala Office, 1997.

Monier-Williams, Monier. *Monier William's Sanskrit-English Dictionary*. 2nd ed. Oxford: Oxford University Press, 1899.

Patnaik, S. and A. Pandey. 'A study of indigenous community based forest management system: Sarna (sacred groves)'. In *Conserving the Sacred* for *Biodiversity Management*. Edited by P.S. Ramakrishnan, K.G. Saxena and U.M. Chandrasekara. New Delhi: Oxford & IBH, 1998.

Prime, Ranchor. *Hinduism and Ecology*. New Delhi: Motilal Banarsidass, 1994.

Ragozin, Z.A. *Vedic India as Embodied Principally in the Rig Veda*. New Delhi: Munshiram Manoharlal, 2005.

Rai, R. 'Sacred Groves in Tribal Pockets of Madhya Pradesh and Plants Conserved by Ethnic Communities in Conservation of Biodiversity'. In *Ecological Traditions of Madhya Pradesh and Chhattisgarh*. Chennai: C.P.R. Environmental Education Centre, 2014.

Rajamani, R. 'Ecological Traditions of Andhra Pradesh'. *Ecological Traditions of Andhra Pradesh*. Edited by Nanditha Krishna. Chennai: C.P.R. Environmental Education Centre, 2005.

Ramanujam, M.P. 'Conservation of environment and human rights: Sacred groves in cultural connections to biodiversity'. *PRP Journal of Human Rights* 4 (2000).

Rath, S.K. 'Narendra Tank in Legend and History'. *Orissa Review*. June 2004.

Rath, S. 'Kautilya's Attitude towards Fauna in the Arthasastra'. In *Kautilya's Arthashastra and Social Welfare*. Edited by V.N. Jha. New Delhi: Sahitya Akademi, 2006.

Renugadevi, R. 'Environmental ethics in the Hindu Vedas and Puranas in India'. *African Journal of History and Culture (AJHC)* 4 (2012). http://www.academicjournals.org/AJHC.

Sanyal, Sanjeev. *Land of the Seven Rivers*. Gurgaon: Penguin Books India, 2012.

Sensarma, P. *Ethnobiological Information in Kautiliya Arthashastra*. Calcutta: Naya Prakash, 1998.

Seshadri, Padma and Padma Malini Sundararaghavan. *It Happened along the Kaveri*. New Delhi: Niyogi Books, 2012.

Shiva, Vandana. *Earth Democracy: Justice, Sustainability and Peace*. London: Zed Books, 2006.

Singh, Govind, Mihir Deb and Chirashree Ghosh. 'Urban Metabolism of River Yamuna in the National Capital Territory of Delhi, India'. *International Journal of Advanced Research* 4 (2016).

Singh, Ritu et al. *Manimahesh Sacred Landscape—A Monograph*. New Delhi: INTACH, 2016.

Sircar, N.N. and R. Sarkar. *Vrksayurveda of Parasara* (A Treatise of Plant Science). New Delhi: Sri Satguru, 1996.

Skolmowski, H. 'Sacred Groves in History'. *Himalaya Man and Nature* XV (1991): 5.

Sontheimer, G.D. 'Hinduism: The Five Components and Their Interaction'. In *Hinduism Reconsidered*. Edited by G.D. Sontheimer and H. Kulke. New Delhi: Manohar, 1989.

Srivastava, A.L. *Life in Sanchi Sculpture*. New Delhi: Abhinav Publications, 1983.

Swami, P. *Encyclopaedic Dictionary of Upanisads*. Vol. 3. New Delhi: Sarup & Sons, 2000.

Sukumaran, S. and A.D.S. Raj. 'Sacred groves as a symbol of sustainable environment—A case study'. In *Sustainable Environment*. Edited by N. Sukumaran. Thirunelveli: SPCES, MS University, Alwarkurichy, 1999.

Tripathi, R.S., B.K. Tiwari and S.K. Barik, *Sacred Groves of Meghalaya: Status and Strategy for Their Conservation* (Shillong: NAEB, NEHU, 1998).

Vaidyanathan, S., and Shayoni Mitra. *Rivers of India*. New Delhi: Niyogi Books, 2011.

Vannucci, M. *Ecological Readings in the Veda*. New Delhi: D.K. Printworld, 1994.

Vartak, V.D. 'Sacred Groves of Tribals for In-situ Conservation of Biodiversity'. In *Ethnobiology in Human Welfare*. Edited by S.K. Jain. New Delhi: Deep Publications, 1996.

Vyas, S.N. *India in the Ramayana Age*. New Delhi: Atma Ram, 1967.